AMAZING SPACEFACTS

SOLAR SYSTEM • STARS • SPACE TRAVEL

Susan Goodman

PETER BEDRICK BOOKS

NEW YORK

First American edition published in 1993 by

Peter Bedrick Books
2112 Broadway
New York, NY 10023

Published by agreement with Oxford University Press,
England.

Library of Congress Cataloging-in-Publication Data
Goodman, Susan.
Amazing spacefacts/Susan Goodman. — 1st American ed.
Indludes index.
ISBN 0-87226-365-7. — ISBN 0-87226-257-X (pbk.)
I. Title.
QB44.2.G66 1993
520—dc20   92-40112 CIP

Designed and typeset by Threefold Design
Printed in Hong Kong
10 9 8 7 6 5 4 3 2 1

► CONTENTS

Photo credits

Carnegie Institute, Washington p9 (right)

G E Astro Space/Science Photo Library p97

David Hardy pp56, 105

Michael Holford pp125, 126, 127, 130, 131

NASA pp15, 28 (left); 30, 31, 32, 34 (left), 34 (right), 35, 37, 38 (top left and right), 38 (bottom left), 30 (bottom right), 40, 43 (top left), 43 (middle left), 43 (bottom left), 46, 57, 107, 111 (top and bottom), 112, 114 (top)

NASA/Science Photo Library pp10, 16, 19, 21, 23, 28 (right), 29 (left and right), 39, 43, (top right), 43 (middle right), 47, 48, 49, 50, 58, 59, 62, 63, 80 (top and bottom), 81 (top and bottom), 94, 99 (bottom), 114 (bottom), 120 (bottom), 138

Science Photo Library/Jodrell Bank p82/Julian Baum pp44, 101/Jean L Charmet p129/Dr Hilmar Duerbeck p85/Jack Finch p11/Hencoup Enterprises p121/David Hardy pp52, 53/Dr Jean Lorre p27 (right)/Mark Paternoster p70/David Parker p135/Roger Rossmeyer (Starlight) pp120 (top), 128/Ronald Royer pp67, 89 (bottom right), 91 (bottom)/John Stanford pp64, 65, 73.

Novosti/Science Photo Library p109 (top and bottom)

Royal Observatory Edinburgh/AATB/Science Photo Library pp88 and 89 (left), 92, 93

Smithsonian Astrophysical Observatory/Science Photo Library p89 (top right, Dr Rudolph Schild)

Tass from Sovfoto p134

X-Ray Astronomy Group, Leicester University/Science Photo Library p90

Yerkes Observatory p118.

Artwork credits

Gary Hincks: 24

Oxford Illustrators: 6, 7, 12, 25, 54, 55, 117

David Hardy: cover, 8, 17, 18, 20, 22, 26, 31, 36, 41, 42, 45, 51, 59, 60, 69, 77, 78, 84, 86, 91, 96, 98, 103

SOLAR SYSTEM

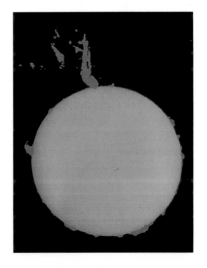

▶ INTRODUCING THE SOLAR SYSTEM

● The Solar System consists of nine major planets in orbit around the Sun, and their moons. There are also asteroids, comets, and the gas and dust particles in the space between them all. In order from the Sun, the major planets are Mercury, Venus, Earth, Mars, Jupiter, Saturn, Uranus, Neptune and Pluto.

● Most of the **asteroids** (also called minor planets), travel round the Sun in a belt between the orbits of Mars and Jupiter. Several thousand are known and there are probably many more. They range in size from about a mile to nearly a thousand miles across.

● Swarms of **meteoroids** also orbit the Sun. These are smaller pieces of rock and dust. Some weigh up to several pounds but most are more like grains of sand. They are invisible from Earth except when they enter the atmosphere. On entering Earth's atmosphere all but the tiniest burn up producing the brief streaks of light we know as meteors, or 'shooting stars'.

● **Comets** travel in long, narrow elliptical orbits around the Sun. As a comet

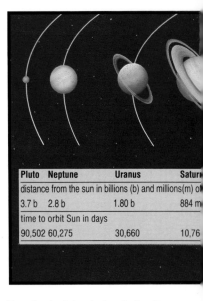

Pluto	Neptune	Uranus	Satur⋯
distance from the sun in billions (b) and millions(m) o⋯			
3.7 b	2.8 b	1.80 b	884 m⋯
time to orbit Sun in days			
90,502	60,275	30,660	10,76⋯

The major planets in order from the Sun. The planets are drawn to scale, but the distances between them are not to scale.

approaches the Sun, it heats up and develops a tail of gas and dust. Some comets take only a few years to complete an orbit of the Sun, but others take tens of thousands of years, or even longer.

● All the planets move round the Sun in the same direction. With the exception of Pluto, their orbits lie almost in the same plane as if resting on a flat surface. So it is quite accurate to show them as eight rings around the Sun on a flat piece of paper. Pluto's orbit is

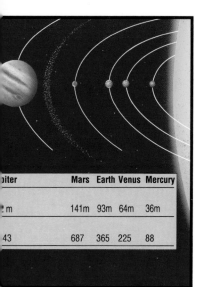

...iter	Mars	Earth	Venus	Mercury
..m	141m	93m	64m	36m
..43	687	365	225	88

tilted, to the main plane of the Solar System, so as Pluto orbits it travels above and below our imaginary flat surface.

● Seven of the nine planets spin in a direction that looks counter-clockwise to someone looking down on the Earth's north pole. On these planets the Sun rises in the east and sets in the west. Earth is one of these planets.
Venus and Uranus are the two planets that spin the opposite way. Uranus is particularly strange because the axis it rotates round is tilted right over. It is rather as if Uranus lies on its side and rolls round the Sun, while the other planets look more like spinning tops in orbit.

● Both Mercury and Pluto have orbits that are elliptical. The paths of the other planetary orbits are almost circular. Pluto's elliptical orbit sometimes brings it inside Neptune's orbit, so it is closer to the Sun than Neptune.

● There are two main kinds of major planet: the rocky ones similar to the Earth, called the **terrestrial planets**, and the gas giants similar to Jupiter, called the **Jovian planets**.

● Earth is a terrestrial planet and so are Mercury, Venus and Mars. Each is basically a solid ball of rock. All are roughly the same kind of size and density.

● Jupiter, Saturn, Uranus and Neptune are the Jovian planets. They are all much larger than the terrestrial planets, and are huge balls of gas and liquid.

● Pluto is not normally included in either group. Little is known about this small, remote planet. It is probably a mixture of rock and ice.

THE SUN

● Our Sun is an ordinary star, just one of the 200 billion stars in the Milky Way Galaxy. It lies 28,000 light years from the center of the Galaxy, in one of the spiral arms.

▶ DID YOU KNOW ?

The Sun travels once around the center of the Galaxy in about 220 million years.

The Sun spins on its rotation axis once every 25 days 9 hours.

● Although it is only a medium-sized star, the Sun is enormous compared to the planets in the Solar system. Its diameter of 1,392,000 km (865,000 miles) is more than a hundred times the Earth's diameter.

● The Sun's mass is about 2,000,000,000,000,000,000,000,000 tons: that is about 99 per cent of the mass of the whole Solar System. About three quarters of it is hydrogen and the rest mostly helium.

● In the core of the Sun, the weight of all the gas pushing down makes it hot and dense enough for hydrogen to be turned into helium by a process called nuclear fusion. Some of the hydrogen is converted entirely into energy.

Cross-section of the Sun.

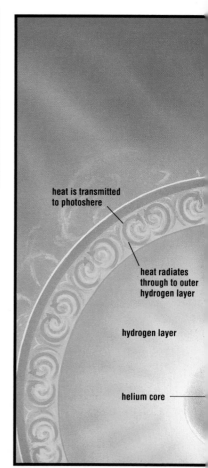

heat is transmitted to photoshere

heat radiates through to outer hydrogen layer

hydrogen layer

helium core

The temperature here is around 59 million° F.

● Energy gradually flows outwards towards the surface of the Sun. In some layers, currents of hot gas rise upwards. They share their heat with the cooler gas above and then sink down again. This process is called convection.

● The yellow surface of the Sun we see is called the photosphere. Its temperature is about 10,000° F. The Sun looks yellow because of its particular temperature: cooler stars are red and hotter stars white.

● Many dramatic events take place on the surface of the Sun. Together these are called 'solar activity'. Solar activity increases and decreases over a cycle lasting about 11 years.

● **Sunspots** are one form of solar activity. These dark blotches can last several months but most disappear in 10 days or less. Sunspots normally occur in groups; individual spots can measure thousands of miles

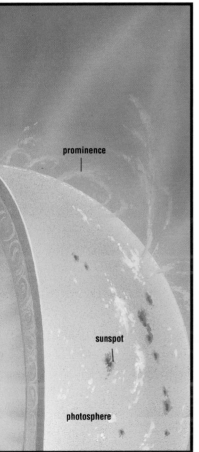

prominence

sunspot

photosphere

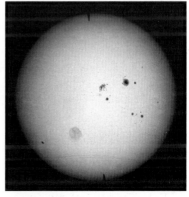

Black and white photo of the Sun showing sunspots.

across. Sunspots look darker only because they are 1,500 degrees cooler than the rest of the photosphere. Seen alone against a dark sky, a sunspot would be as bright as the full Moon! The cooling effect is caused by the strong magnetic field in a sunspot.

● Solar flares are another kind of solar activity. These often occur near sunspots. In a flare, a huge amount of energy is released in just a few minutes and the temperature can go up to more than 200 million° F. Flares produce bursts of X-rays, radio waves and clouds of atomic particles that stream into space. When these particles enter the Earth's

Spectacular eruptions of hot gas leap upwards from the Sun's surface, shooting into space at speeds of hundreds of miles per second. These are called prominences. Typically they last for a few hours, though there is a kind that can last for weeks.

The northern lights, aurora borealis, photographed in Alaska. Auroras are caused by charged subatomic particles from the Sun, interacting with atoms, and molecules in the Earth's atmosphere, above about 60 miles. The particles from the Sun are concentrated over the poles by the Earth's magnetic field.

atmosphere they cause bright lights in the sky called the northern and southern lights (the aurora) sometimes seen in the night sky, usually in the polar regions. They can also upset the Earth's own magnetism and affect long-distance radio communications. The atmosphere protects us from the harmful effects of solar X-rays but, when astronauts were visiting the Moon, a careful watch was kept for flares.

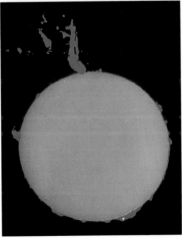

False-color photograph of solar flares, and a large solar prominence (top).

Astronauts have to return to their spacecraft for protection as the Moon has no atmosphere.

▶ SOLAR ECLIPSES

● Although the Sun is 400 times larger than the Moon, it appears the same size in the sky because it is 400 times further away.

● When the Moon lies directly between the Earth and the Sun, it blocks the Sun's light and, from some parts of the Earth, a **total solar eclipse** is visible. During a total eclipse you cannot see the Sun and everything is dark.

There is an eclipse of the Sun when the Sun, Moon and Earth line up, and the Moon's shadow falls on the Earth. (Not drawn to scale.)

● A partial eclipse occurs when the Moon does not quite cross directly in front of the Sun. Part of the Sun is visible so there is still some light.

● The distances between the Earth, Moon and Sun vary to some extent because the Moon's orbit round the Earth and the Earth's orbit round the Sun are elliptical, and not perfect circles. This is called an **annular eclipse** from the Latin word *annulus*, meaning ring.

● During a total eclipse of the Sun, the Moon blocks out the blinding light of the Sun's surface and the faint, outer parts of the Sun become visible.

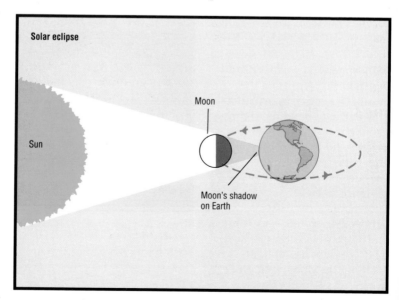

Solar eclipse

Sun

Moon

Moon's shadow on Earth

Solar eclipses			
Date		Type	Region of visibility
1992	June 30 Dec 24	total partial	South Atlantic Ocean Arctic
1993	May 21 Nov 13	partial partial	Arctic Antarctic
1994	May 10 Nov 3	annular total	Pacific Ocean, Mexico, USA, Canada, Atlantic Ocean, Peru, Brazil, South Atlantic Ocean
1995	Apr 29 Oct 24	annular total	South Pacific Ocean, Peru, Brazil, Iran, India, East Indies, Pacific Ocean

The solar corona appears as a milky white halo around the Sun during total eclipse. The corona extends outwards for many times the Sun's diameter. It contains particles at temperatures of millions of degrees.

THE SUN'S FUTURE

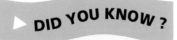

DID YOU KNOW ?

Every second, the Sun uses up 400 million tons of hydrogen completely and changes it into energy.

When will the Sun exhaust its supply? Scientists estimate that the Sun's total lifetime is about 10 billion years. About half of this has gone by already. When the supply of hydrogen in its core does begin to run out, the Sun will gradually expand to be about a hundred times bigger than it is now. It will engulf the Earth in the process. The Sun's surface temperature will fall and the color will change to red. The Sun will become a type of star known as a red giant. Ultimately, this red giant will shrink again to become a white dwarf, about the size of the Earth.

▶ THE PLANETS

Comparing the planets

Planet	Diameter (Earth=1)	Mass (Earth=1)	Density (g/cm³)	Orbit period round Sun (Earth days)	Average orbital velocity (miles per second)
Mercury	0·38	0·06	5·4	88	29.7
Venus	0·95	0·82	5·3	225	21.7
Earth	1·00	1·00	5·5	365	18.5
Mars	0·53	0·11	3·9	687	15
Jupiter	11·21	316·9	1·3	4,333	8
Saturn	9·45	95·2	0·7	10,759	6
Uranus	4·01	14·5	1·3	30,685	4
Neptune	3·88	17·1	1·6	60,190	3.3
Pluto	0·18	0·0022	2·0	90,800	3

The masses and diameters are given relative to the Earth's to make it easy to compare them with the Earth.
The diameters given are those at the equator. The Jovian (gas) planets bulge at their equators because they are spinning rapidly.
The orbital periods are given to the nearest day.

MERCURY

● The images returned by the space probe *Mariner 10* in 1974 and 1975, revealed that Mercury has a cratered, rocky surface, rather like the Moon's, but without the larger, smoother areas. It has an enormous crater, the Caloris Basin, which is over 1,300 km (813 miles) across.

● Until 1974 almost nothing was known about the surface of Mercury. Whenever it is visible from Earth it is never far from the Sun, low down in the sky. That makes it very difficult to get a good view of Mercury in a telescope.

A photomosaic of Mercury using pictures taken by *Mariner 10*.

● In 1974 the American space probe *Mariner 10* passed within a few hundred miles of the surface of Mercury and relayed back to Earth thousands of images. This first flyby was in March 1974. It was followed by two more, in September 1974 and March 1975. The images returned during these encounters have been used to map about 35 per cent of Mercury's surface.

● Mercury is the second smallest planet after Pluto, which is half its size. It is the planet nearest the Sun, orbiting faster than any other. A complete journey round the Sun takes Mercury just 88 days at an average speed of about 172,000 km/h (107,000 mph). Earth travels at about two-

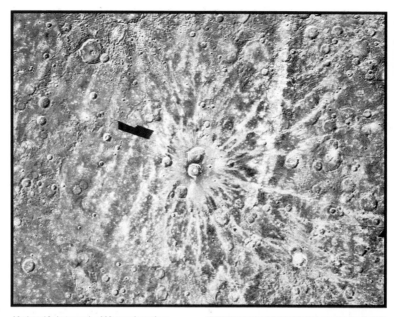

Mariner 10 photograph of Mercury's surface showing the impact crater Degas. The black region is an area not photographed by the spacecraft.

thirds this speed and Pluto at only about one tenth.

● Astronomers think Mercury has an iron core accounting for about 70 per cent of its mass and 75 per cent of its diameter. This core is a little bigger than our Moon.

● The mantle surrounding the core is thought to be a layer of rock about 640 km (400 miles) thick.

● *Mariner 10* detected atoms of sodium and tiny amounts of helium and hydrogen near the

Mercury fact box	
Average distance from Sun	58 million km (36 million miles)
Length of year	88 Earth days
Length of solar day	156 Earth days
Rotation period relative to stars	58·65 Earth days
Diameter at equator	4,870 km (3,030 miles)
Approximate surface temperatures	equator, midday: 420°C (800°F); midnight: -180°C (-300°F)

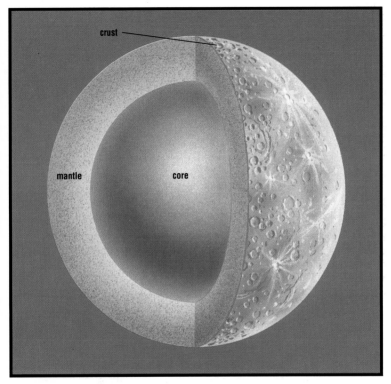

Cross-section of the planet Mercury.

surface of Mercury, but this gas can hardly be called an atmosphere: the best vacuum that can be made in a laboratory on Earth has more gas in it!

● During a day on Mercury, Mercury's distance from the Sun varies between 70 million km (43 million miles) and 56 million km (35 million miles). An observer on Mercury would see the Sun increase in size and then decrease again.

▶ **DID YOU KNOW ?**

A solar day on Mercury, from sunrise to sunrise, lasts about six Earth months.

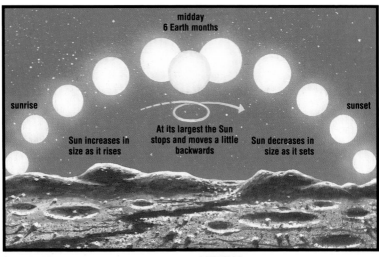

midday
6 Earth months

sunrise

sunset

Sun increases in
size as it rises

At its largest the Sun
stops and moves a little
backwards

Sun decreases in
size as it sets

A day on Mercury (sunrise to sunrise) is almost
two Mercurian years long. During a day Mercury
moves nearer the Sun, so the Sun seems to
increase in size, and then shrink again.

● At midday on Mercury, with
the Sun overhead, the
temperature reaches around
420°C (800°F). Three Earth
months after sunrise, the Sun
sets and the cold night begins.
With no atmosphere to act like
a blanket and hold in the
warmth, the temperature
plunges to -180°C (-300°F).

VENUS

Venus fact box	
Average distance from Sun	108 million km (67 million miles)
Length of year	225 Earth days
Length of solar day	584 Earth days
Rotation period relative to stars	243 Earth days
Diameter at equator	12,100 km (7,500 miles)
Approximate surface temperature	430 - 480°C (810 - 900°F)

● Venus is only slightly
smaller than Earth, and comes
nearer to us than any other

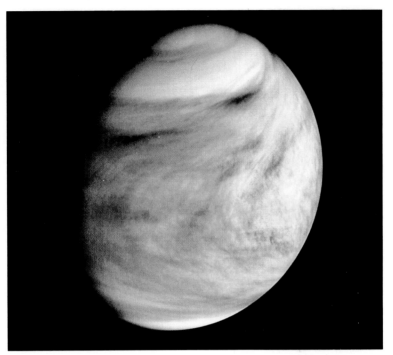

Venus, the second major planet from the Sun, and the one that comes closest to Earth. It has a yellow-white cloud cover which permanently obscures the surface.

planet. At its closest it is about 42 million km (26 million miles) away. More space probes have been sent to Venus than to any other planet, because it is so close.

● Venus is shrouded by a dense blanket of poisonous gases. Clouds of sulphuric acid prevent us from seeing the surface of the planet through telescopes, but space probes have used radar to map the

▶ **DID YOU KNOW ?**

The probes sent to land on the surface of Venus had to be able to withstand temperatures of almost 500°C (900°F) (a temperature high enough to melt tin) and atmospheric pressure ninety times greater than that at the surface of the Earth.

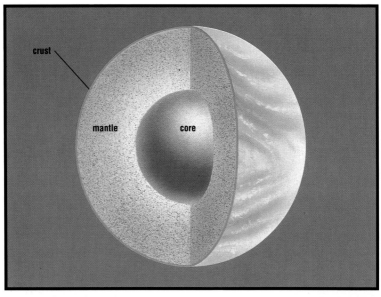

Cross-section of the planet Venus.

planet surface revealing mountains and highland areas surrounded by large, flat deserts of broken and eroded rocks.

● The gas clouds reflect sunlight so well, Venus appears in our sky as a brilliant object; only the Sun and Moon are brighter. Because Venus lies between the Earth and the Sun, like Mercury it is always seen fairly close to the Sun in our sky. Depending on whether it is visible just before sunrise or just after sunset, it is sometimes described as the 'morning star', or the 'evening star' (though it is, of course, a planet and not a star at all).

● Astronomers think Venus has a core containing iron, but the details of what it might be like are not known. The core is surrounded by a rocky mantle.

● The surface has been photographed by Soviet space probes that landed on Venus, and a thimbleful of Venusian soil was automatically analyzed by one of them. The sample was found to contain calcium, sodium, silicon, iron, magnesium and potassium.

● The atmosphere of Venus consists of carbon dioxide gas

Space probes to Venus

1967	*Venera 4* (USSR)	Explored atmosphere of Venus. Attempts to land *Venera*s *4*, *5* and *6* failed because they did not withstand crushing by the atmosphere.
1970	*Venera 7* (USSR)	First space probe to land on Venus.
1972	*Venera 8* (USSR)	Landed on surface and transmitted data back to Earth.
1974	*Mariner 10* (USA)	First TV images from flyby.
1975	*Venera 9* (USSR)	First pictures from surface.
1978-80	*Pioneer Venus 2* (USA)	Mapped almost whole surface.
1982	*Venera 13* (USSR)	Landed in volcanic highland region. Returned first color images of surface and carried out soil analysis.
1982	*Venera 14* (USSR)	Landed 5 days after *Venera 13* in lowland region.
1990	*Magellan* (USA)	In August, arrived in orbit around Venus and began returning detailed radar images of planet's surface

containing clouds of sulphuric acid. The atmosphere is about 250 km (150 miles) deep, but most of the gas is concentrated in the lowest 28 km (17 miles).

● The carbon dioxide acts like a greenhouse, trapping the Sun's heat so that the surface temperature reaches 480°C (900°F) and never falls below 430°C (810°F).

False-color view of the surface of Venus, showing the western Eistla Region. In the foreground you can see a large rift valley, and the mountain, top right, is 2 miles in height.

EARTH

● Viewed from space, planet Earth has blue oceans, white clouds and polar ice caps, and dark areas of land. It is the only planet we know that has life. Life on Earth only thrives because there is a balance between animals, which use up oxygen and release carbon dioxide; and plants, which take in carbon dioxide and produce oxygen. For all life to survive, a steady temperature is important: a few degrees hotter and the ice caps will melt, flooding large areas of land; a few degrees colder and crops will not grow.

Cross-section of the Earth.

The core is at about 4,000°C (7,200°F) and mainly consists of iron and nickel which are solid in the inner part of the core and liquid in the outer part.

The mantle of molten rock ranges in temperature between 1,500 and 3,000°C (2,700 and 5,400°F).

The crust is about 10 km (6 miles) thick under the oceans and 30 km (20 miles) thick where there are continents. The average temperature is about 22°C (72°F).

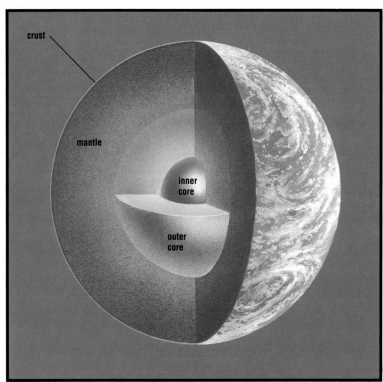

crust

mantle

inner core

outer core

The Earth, taken from *Apollo 17* in 1972, and showing Antarctica, Africa and Arabia.

● The Earth's crust contains over 2,000 different minerals. Only about 20 of these minerals account for over 90 per cent of the crust. Minerals are made up of different chemical elements. Just eight elements make up 99 per cent of the Earth's crust.

We need to be very careful not to disturb this delicate balance and turn our living planet into a barren desert such as Mars or Venus.

Element	% by weight
Oxygen	47
Silicon	28
Aluminium	8
Iron	5
Calcium	3½
Sodium	3
Potassium	2½
Magnesium	2

Earth fact box	
Average distance from Sun	150 million km (93 million miles)
Length of year	365 Earth days
Length of solar day	24 hours
Rotation period relative to stars	23 hours 56 minutes 4 seconds
Diameter at equator	12,760 km (7,930 miles)
Average surface temperature	22°C (72°F)

● Area of land: 148,326,000 km² (57,268,900 square miles), equals 29 per cent of the total surface of the Earth.

● Area of water: 361,740,000 km² (139,668,500 square miles), equals 71 per cent of the total surface of the Earth.

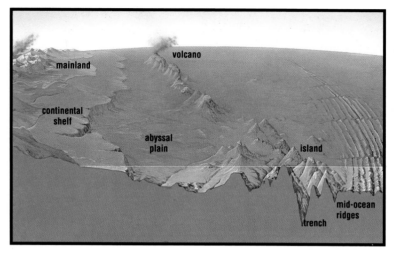

mainland

volcano

continental
shelf

abyssal
plain

island

mid-ocean
ridges

trench

Over 70% of the Earth's surface is underwater.

● Highest mountain: Mount Everest, 8,848 m (29,000 ft) above sea level.

● Deepest ocean trench: Mariana Trench, Pacific Ocean, 11 km (36,000 ft) below sea level.

There are about 700 active volcanoes on Earth, including some underwater.

● There are about 700 active volcanoes. About 30 of them erupt each year.

● About a fifth of the Earth's land surface is mountainous and a further eighth is covered by deserts.

● Earth is the only planet in the Solar System with liquid

The structure of Earth's atmosphere. The heights given for the different layers are not exact as they merge into one another, and can vary depending on time of year, latitude, and solar activity.

water. Of all the water on Earth, 97 per cent is saltwater. Only 3 per cent is freshwater and two-thirds of this is frozen into ice. Only 1 per cent of all the Earth's water is in lakes, rivers or underground.

▶ DID YOU KNOW ?

Ice permanently covers about 10 per cent of the Earth's surface.

● The largest freshwater lake is Lake Superior on the border between Canada and the USA. Its area is 82,400 km^2 (31,800 square miles). The Caspian Sea is the largest saltwater lake, or inland sea. It is about 372,000 km^2 (144,000 square miles) in area.

● The oxygen in the Earth's atmosphere is produced by plants. About 70 per cent of it comes from tiny plants, called algae, floating in the oceans. Plants also take in the carbon dioxide that animals breathe out.

MARS

Mars fact box	
Average distance from Sun	228 million km (142 million miles)
Length of year	687 Earth days
Length of solar day	24 hours 39 minutes 35 seconds
Rotation period relative to stars	24 hours 37 minutes 23 seconds
Diameter at equator	6,670 km (4,150 miles)
Average surface temperature	-23°C (-10°F) (-150°C (-240°F) at poles, 0°C (32°F) at equator)

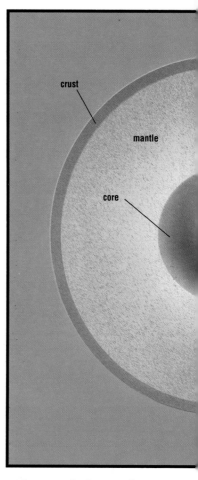

● The popular name for Mars is 'the Red Planet' because of its obvious color when it appears in the sky. The surface rocks are colored by a large amount of the chemical iron oxide, which most people are familiar with as ordinary rust. Even the sky on Mars looks pink, due to the dust blown up into the atmosphere.

● Before space probes were sent to Mars, the best Earth-based telescope revealed only that the surface was reddish, with some darker markings, and white ice caps at the north and south poles.

● Mars rotates round an axis tilted by 24°. This is not very different from the Earth's, which is at 23½°. As a result, Mars goes through seasons

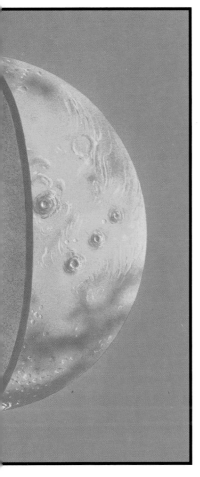

Cross-section of the planet Mars.

Viking orbiter's color enhanced photograph of half the planet Mars, showing both polar ice caps.

the southern one at its smallest. After half a Martian year, the situation is reversed. The polar caps are a thin layer of frozen carbon dioxide and water.

● The only water now on Mars seems to be frozen, but images of the Martian landscape show channels and canyons that were obviously cut by water flowing in the past.

● Mars is thought to have a core containing iron, but its precise size and composition are not known. The core is surrounded by a rocky mantle.

● The two *Viking* spacecraft that landed on Mars in 1976 analyzed samples of soil to see what chemicals they contained. They found that oxygen combined with silicon makes

very similar to the ones we have on Earth. When it is winter in the northern hemisphere and summer in the southern hemisphere, the north polar cap is at its largest and

The Martian surface and sky taken by *Viking 1*. Part of the spacecraft can be seen in the corner.

● Images of the surface near the *Viking* landing places show a rocky desert landscape. Mars

up 44 per cent of the soils and iron oxide 19 per cent. Small quantities of sulphur, magnesium, aluminium, calcium and titanium were also detected.

View of Mars showing the giant volcano Olympus Mons, top right. The Tharsis mountains are center right.

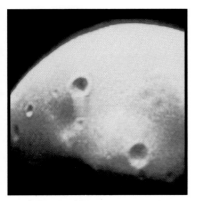

This photograph of Deimos was taken by the *Viking Orbiter 1*. The craters you can see measure up to a mile in diameter.

Phobos is the largest of Mars' two satellites. It has a huge crater called Stickney, which is 5 miles across.

has several very large, extinct volcanoes and a giant canyon. Olympus Mons is the largest of the volcanoes. It is over 25 km (15 miles) high and has a crater more than 80 km (50 miles) across.

● Mars has two tiny moons. Both are roughly potato-shaped and have craters on them.

● **Deimos** is about 10 miles long and 7½ miles across. It orbits Mars at a distance of 23,500 km (14,500 miles), taking 1¼ Earth days.

● **Phobos** is about 28 km (17 miles) long and 20 km (12 miles) across. It orbits Mars at a distance of 9,340 km in just under 8 (Earth) hours.

Space probes to Mars		
1965	*Mariner 4* (USA)	First successful flyby
1971	*Mariner 9* (USA)	First probe to orbit Mars. Returned thousands of images of the surface.
1976	*Vikings 1 & 2* (USA)	First successful landings, together with orbiters. Collected wide range of data and images.

JUPITER

Jupiter fact box	
Average distance from Sun	778 million km (483 million miles)
Length of year	11·9 Earth years
Average rotation period*	9 hours 50 minutes
Diameter at equator	143,760 km (89,350 miles)
Approximate surface temperature	-240°F

*Because Jupiter is not solid, different parts rotate at different rates.

▶ DID YOU KNOW ?

Jupiter is the largest planet of the Solar System: its mass is two and a half times the mass of all the other planets put together. Its diameter is eleven times the Earth's, and its volume 1,300 times greater.

● This giant planet spins faster than any other. At the equator, the velocity due to the spinning motion is about 45,500 km/h (28,400 mph). The equivalent velocity on Earth is about 420 km/h (320 mph). Jupiter bulges at its equator because it is not solid, and is spinning so fast.

Jupiter and two of its satellites (Io, left, and Europa) taken by *Voyager 1* in 1979.

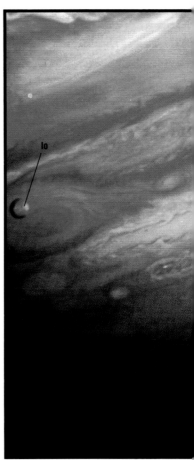

Io

● Jupiter has no definite surface but, under the outer cloud layers, the gas thickens as the pressure gets greater, gradually becoming like a hot liquid. The atmosphere is stormy, with high winds and lightning, which is probably quite violent.

● Three faint rings of fine particles circle Jupiter. They are only a few miles thick and cannot be seen from Earth even with the most powerful

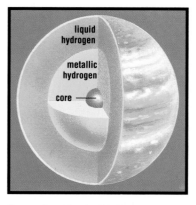

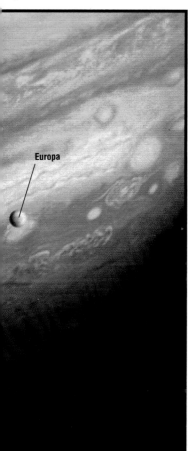

Europa

Cross-section of the planet Jupiter.

telescopes. They were first discovered on images returned by the spacecraft *Voyager 1*. The rings form a band about 50,000 km (31,000 miles) above Jupiter's cloud tops.

● At its center, Jupiter almost certainly has a small rocky core. The temperature there is around 36,000 - 54,000°F.

● In the layer over the core, the very high pressure makes the liquid hydrogen behave like a molten metal.

● Jupiter is mainly made of hydrogen (88 per cent) and

helium (11 per cent). Small amounts of other chemicals give the clouds their variety of colors.

Jupiter's Great Red spot: the white cloud next to it revolved about its center for six days before disappearing.

● Jupiter's Great Red Spot is thought to be a swirling storm cloud currently about 26,000 km (16,000 miles) long and 12,000 km (7,500 miles) wide. Since it was first observed in 1664, its length and color have varied. Why it is red is still a mystery. Small white clouds that get caught up around the swirling Red Spot take about 7 days to go right round it and then they merge in with the cloud bands.

Space probes to Jupiter (all US)

1973	*Pioneer 10*	First flyby of Jupiter
1974	*Pioneer 11*	Flyby
1974 March	*Voyager 1*	Close flyby returning detailed images of planet and five moons
1979 July	*Voyager 2*	Similar mission to *Voyager 1*

● Four of Jupiter's moons were discovered in 1610 by the Italian astronomer Galileo, when he first used a telescope; so they are known as the Galilean satellites. None is more than a tiny spot of light when viewed with even the most powerful telescopes. Images sent back by the *Voyager* spacecraft showed just how different they are from each other.

Galilean Moons of Jupiter

Name	Diameter (miles)	Average distance from planet (miles)	Description
Io	2,250	261,400	Surface colored due to molten sulphur thrown out from active volcanic centers
Europa	1,945	416,000	Smooth ice-covered surface criss-crossed by network of dark lines
Ganymede	3,262	663,400	Largest moon in the Solar System, has dark cratered areas and lighter grooved terrain
Callisto	2,976	1,167,460	Very heavily cratered

The surface of Io showing volcanic activity, taken by *Voyager 1*.

Europa, the surface facing away from Jupiter.

SATURN

Saturn fact box	
Average distance from Sun	1,427 million km (887 million miles)
Length of year	29·5 Earth years
Average rotation period*	10 hours 32 minutes
Diameter at equator	120,420 km (74,840 miles)
Approximate surface temperature	-300°F

*Because Saturn is not solid, different parts rotate at different rates.

Space probes to Saturn (all US)		
1979	*Pioneer 11*	First flyby
1980	*Voyager 1*	Close flyby with detailed images of planet, rings and moons returned
1981	*Voyager 2*	Similar mission to Voyager 1

● Saturn, like the other gas giants, is made mainly of hydrogen and helium. It has the lowest density of all the planets, only seven tenths that of water.

This image taken by *Voyager 2* shows the rings of Saturn. The different colors reveal the presence of different kinds of materials. The colors have been exaggerated by processing in a computer to show up the differences more clearly.

DID YOU KNOW ?

Saturn has a density lower than water, so if there was a tub of water big enough, Saturn would float in it.

● Astronomers think Saturn has a small rocky core. Over the core is a thick layer of liquid hydrogen. The deepest parts of this layer are under so much pressure that the hydrogen acts like a molten

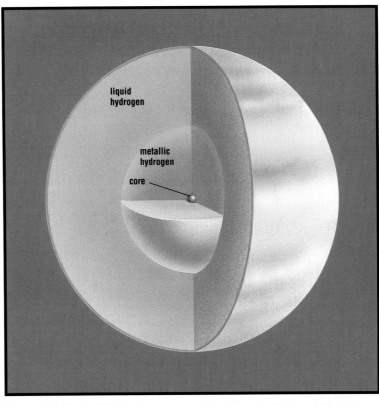

liquid
hydrogen

metallic
hydrogen

core

Cross-section of the planet Saturn.

metal. Like on Jupiter, the
outer cloud layers gradually
merge with the liquid
underneath, and there is no
definite surface.

● Viewed from Earth, three
main rings can be seen around
Saturn, separated by dark gaps.
The *Voyager* spacecraft
identified four other fainter
rings, one inside and three
outside the three brightest. In
the *Voyager* images, the rings
look 'grooved', as if they are
made up of large numbers of
narrow ringlets. The rings
consist of many individual
pieces of a mixture of ice and
dust, ranging in size from a
few inches to a few
yards across.

Voyager 2 photograph of Saturn, taken 21 million miles from the planet. The Moons Rhea and Dione appear as blue dots to the south and southeast of Saturn respectively.

SATURN'S MOONS

● Astronomers using Earth-based telescopes have identified eleven moons orbiting Saturn, and the *Voyager* spacecraft discovered seven more. However, the exact number of moons is unknown. There are probably more of them, in the form of rocks no larger than a few miles across.

● Saturn's largest moon, Titan, is particularly interesting: it is the only moon in the Solar System with a thick atmosphere. Titan appears orange in color, but its opaque atmosphere prevents us seeing the surface. US space scientists have planned a space probe that they hope will land on Titan in about 2002.

Saturn's largest moons

Name	Diameter (miles)	Average distance from planet (miles)
Mimas	242	115,022
Enceladus	310	147,572
Tethys	650	182,689
Dione	694	233,988
Rhea	947	332,964
Titan	3193	757,547
Hyperion	180	918,220
Iapetus	893	2,208,000
Phoebe	136	8,030,240

Titan

Mimas

Enceladus

Iapetus

URANUS

Uranus fact box	
Average distance from Sun	2,869 million km (1.783 billion miles)
Length of year	84 Earth years
Average rotation period	17 hours 15 minutes
Diameter at equator	51,120 km (31,770 miles)
Approximate surface temperature	-410°F

● Uranus was the first planet to be discovered with the help of a telescope. (All the others are bright enough to be seen without one.)

▶ **DID YOU KNOW ?**

Uranus was found in 1781 by the astronomer Sir William Herschel.

Voyager 2 sent back the first close-up view of Uranus in January 1986, showing a bluish disc with no obvious features. Sunlight scattered from the top layers of Uranus's atmosphere made it impossible to see anything underneath. This is a composite photo of Uranus (center), with moons.

False-color view of the rings of Uranus, taken by *Voyager 2*. All nine known rings are visible.

● Nearly 200 years after its discovery, in 1977, astronomers discovered that Uranus has a system of rings around it. They are too faint to be seen directly from Earth, even with a powerful telescope but, in 1977, Uranus's path through the sky took it directly in front of a star. The starlight was observed to dim and brighten again several times before the star disappeared behind the planet. The same thing happened just after the star reappeared. A set of rings around Uranus was the only sensible explanation. The existence of eleven separate rings was confirmed when *Voyager 2* passed only 75,000 km (46,600 miles) away from Uranus in 1986. They are made of pieces of rock as dark as coal.

● Uranus probably has a small rocky core but the bulk of the planet is a mantle made of rock, water, methane and ammonia. Deep inside the planet, the mixture will be a hot liquid, but near the surface it is frozen. The outermost layer is an atmosphere, mainly of hydrogen, with helium and small amounts of other chemicals, including methane.

● The axis Uranus spins around is tilted right over to lie nearly in the main plane of the Solar System. This makes Uranus unique among the planets: all the others spin with their axes at right angles to their orbits, or not far off. Uranus appears to roll around on its side as it orbits the Sun, with its spin axis pointing towards the Sun. This means that one half of Uranus faces the Sun for many Earth years. Days and nights are 42 Earth years long on some parts of the planet.

Cross-section of the planet Uranus.

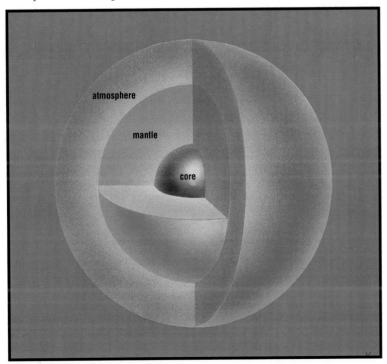

atmosphere

mantle

core

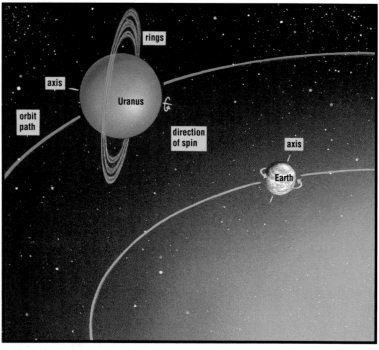

Uranus, unlike the other planets in the Solar System, spins around an axis that is tilted so much it almost lies flat on its plane of orbit.

Uranus's largest moons		
Name	Diameter (miles)	Average distance from planet (miles)
Miranda	290	80,463
Ariel	720	118,568
Umbriel	725	164,900
Titania	980	270,220
Oberon	942	361,212

Oberon

Miranda close up

Umbriel

Ariel

In 1986 *Voyager 2* flew close enough to Uranus to return quite detailed images of five of Uranus's moons. From Earth these are no more than pinpoints of light. It also discovered ten more small moons.

Titania

NEPTUNE

Neptune fact box	
Average distance from Sun	4,497 million km (2. 794 billion miles)
Length of year	165 Earth years
Average rotation period*	19 hours 6 minutes
Diameter at equator	49,530 km (30,780 miles)
Approximate surface temperature	-410°F

*Because Neptune isn't solid, different parts rotate at different rates.

● After Uranus was discovered, astronomers were able to follow its motion through the sky, calculate its orbit round the Sun and predict where it would be in the future. Some years later, it was clear that Uranus was not keeping to the path they expected. Two mathematicians, John Couch Adams in England and Urbain Le Verrier in France, independently worked out that the gravity of another, unknown planet could be pulling Uranus off course, and

Artist's impression of the planet Neptune as seen from Triton, its largest moon.

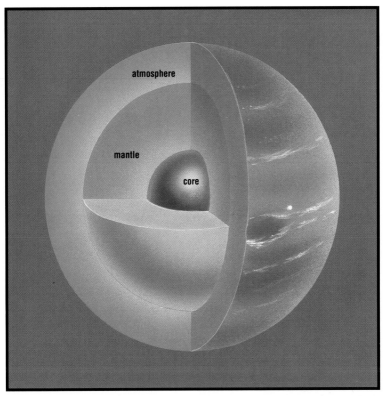

Cross-section of the planet Neptune.

they predicted where this planet might be found in the sky. In September 1846, John Galle and Heinrich D'Arrest, who worked at the Berlin Observatory, turned their telescope towards the predicted position and there was the missing planet, Neptune.

● Even with today's powerful telescopes, Neptune appears only as a faint blue disc. In August 1989 *Voyager 2* flew past Neptune, returning images that reveal details of its atmosphere. In many ways, Neptune is very like Uranus and astronomers think it is likely to have a similar structure.

● Neptune probably has a rocky core about the size of the

Earth. The mantle is a mixture of partly frozen water, ammonia and methane. It is surrounded by a dense atmosphere of hydrogen and helium extending about 9,000 km (5,600 miles) above the planet's icy surface.

● *Voyager 2* images of Neptune captured in August 1989 reveal bands of different shades of blue. The color is due to methane in the atmosphere. A large dark oval patch, named the Great Dark Spot, is thought to be an immense circulating storm, rather like the Great Red Spot on Jupiter.

NEPTUNE'S MOONS

● *Voyager 2* discovered six previously unknown moons

Neptune, showing its Great Dark Spot (left) and Dark Spot 2 below.

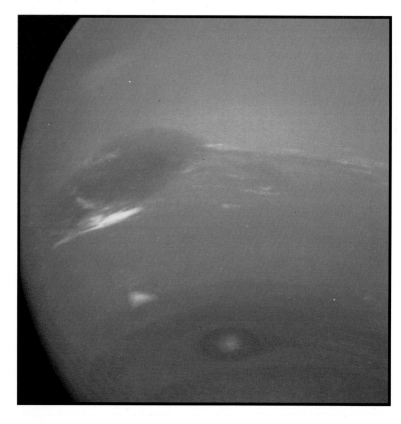

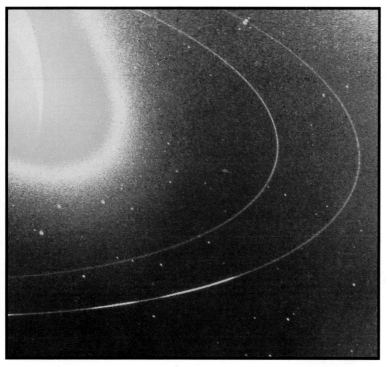

The planet Neptune's two major rings. The radius of the inner ring is 32,860 miles, and that of the outer ring 39,060 miles. They are composed of microscopic dust particles.

rather unusual: all the other larger moons in the Solar System go round the other way.

orbiting Neptune. Just two were already known from Earth-based observations: Triton and Nereid.

● Triton is the largest of Neptune's moons. Its diameter is about 2,700 km (1,680 miles) and it orbits Neptune at a distance of 355,000 km (221,000 miles) every 5·9 days, traveling from east to west. This makes it

▶ **DID YOU KNOW ?**

Triton is thought to be one of the coldest bodies in the Solar System, with a surface temperature of about -455°C.

Color-enhanced *Voyager 2* image of Neptune's largest moon, Triton. The large south polar cap which fills most of the image is thought to consist of a layer of frozen nitrogen. No large impact craters can be seen on the surface of Triton which leads astronomer's to believe that Triton's crust has been renewed relatively recently, that is within the past billion years.

● Nereid is a much smaller moon than Triton. It has an extraordinary elliptical orbit that brings it within 140,000 km (87,000 miles) of Neptune, and also takes it almost ten million kilometres (6 million miles) away. It takes 360 Earth days for Nereid to complete an orbit of Neptune.

PLUTO

Pluto fact box	
Average distance from Sun	5,913 million km (3.675 billion miles)
Length of year	249 Earth years
Average rotation period*	6 days 9 hours 17 minutes
Diameter at equator	2,280 km (1,430 miles)
Approximate surface temperature	-446°F

*Because Pluto is not solid, different parts rotate at different rates.

● Pluto's existence was predicted in the early twentieth century as astronomers tried to find a reason for the orbit of Uranus not following the path they had predicted. Neptune alone could not account for all the discrepancies in the orbit of Uranus. The American astronomer Clyde Tombaugh spent thousands of hours searching photographs with millions of stars, before he found Pluto in 1930. It appears only as a speck even in the largest telescopes, and has not been visited by any space

Pluto as photographed by the Hubble space telescope.

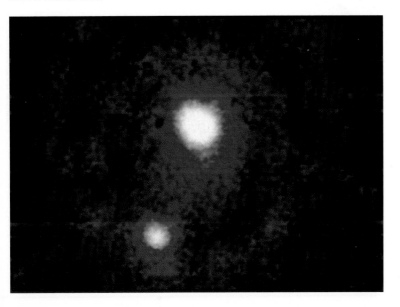

probes, so astronomers have had to work out as best they can what Pluto might be like. Observations from Earth show that its surface is covered with frozen water and methane at about -446°F.

● Pluto has one moon, Charon, which was discovered in 1978 as a small bulge in an image of Pluto, made by J. Christy, an astronomer at the United States Naval Observatory. It is about 1,190 km (740 miles) in diameter, about half the size of Pluto itself. Its surface is icy but

The planet Pluto, the ninth and smallest planet of our solar system, with its moon Charon in the background.

darker than Pluto's, and it seems to have water ice on it but no methane.

● In 1987 and 1988, Charon's orbit carried it behind and directly in front of Pluto on each circuit, as seen from Earth. This is a rare series of events that can only be witnessed every 200 years. Observations made at this time were a great help to astronomers, in working out the size and nature of this remote pair.

● Pluto rotates once on its axis, in exactly the same time that it takes Charon to complete one orbit: 6 days 9 hours and 17

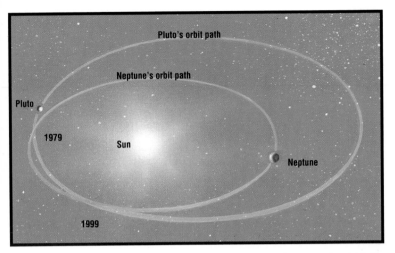

Pluto's orbit path

Neptune's orbit path

Pluto

1979

Sun

Neptune

1999

minutes. This means that Charon is always in the same place in the sky above Pluto.

● Pluto is usually described as the planet furthest from the Sun, but this is not always true. Its orbit is such a stretched-out ellipse that the planet comes to within 2.8 million miles of the Sun, closer than the orbit of Neptune. Its furthest point from the Sun is over 7 million kilometres (4 million miles) away.

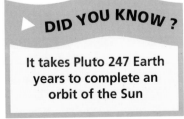

DID YOU KNOW ?

It takes Pluto 247 Earth years to complete an orbit of the Sun

Pluto's orbit is an ellipse, so that although it is the planet furthest from the Sun most of the time, it sometimes passes within Neptune's orbit.

● For twenty years of each orbit Pluto is closer to the Sun than Neptune. During the years 1979 to 1999, Neptune is the outermost planet of the Solar System.

PLANET X?

● The existence of Neptune, and then of Pluto, was predicted to explain why Uranus was deviating from its expected course. Some astronomers do not think that was the end of the story. Pluto does not seem to be big enough to have much effect on Uranus's orbit, and some people believe there is another planet, Planet X, waiting to be discovered.

▶ THE MOON

● The Earth has just one natural satellite, the Moon. At night the Moon is the brightest thing that can be seen in the sky but it does not produce any light of its own: it only reflects sunlight.

● The Moon travels once round the Earth in 27·3 days. Over this time, it also spins once on its axis. As a result, we always see the same side of the Moon. Photographs of the Moon's far side were first seen in 1959 when the Soviet space probe *Luna 3* orbited the Moon and transmitted pictures back to Earth. These first

This photograph of the far side of the Moon, taken by a Lunar Orbiter, shows that is not exactly like the side we always see. It is more evenly cratered and there are hardly any of the darker plains known as maria ('seas'). The purple dots are the crash-landing sites of the probe *Orbiter*.

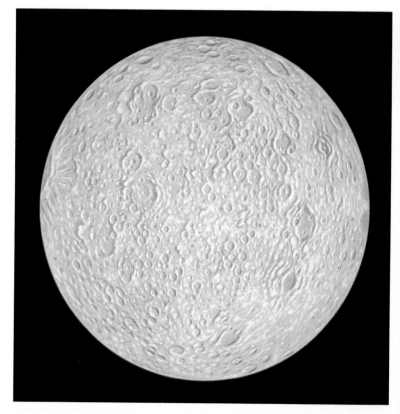

photographs were of rather poor quality but, between 1966 and 1967, the United States sent a series of five Lunar Orbiters, which took thousands of photographs of the Moon's surface.

PHASES OF THE MOON

● Only the half of the Moon facing towards the Sun is lit up: the rest is dark. As the Moon travels around the Earth, the proportion of the bright side visible to us changes. When the Moon is between the Earth and the Sun, we cannot see the lit side at all. This totally dark phase of the Moon is called New Moon. Full Moon is seen when the Earth is between the Sun and the Moon.

The near side of the Moon showing landing sites. Five different missions visited the Moon: *Luna* (pink), *Ranger* (orange-red), *Surveyor* (green), *Orbiter* (purple), and *Apollo* (blue). Only *Apollo* was manned. The dark plains are 'maria'.

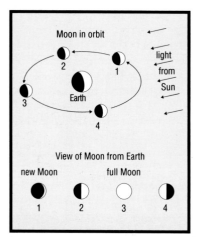

The phases of the Moon. The Sun lights up the half of the Moon facing it. In four weeks, the Moon travels right around the Earth in its orbit. You see different amounts of the lighted half as the moon moves.

Moon fact box	
Average distance from Earth	384,400 km (238,900 miles)
Lunar month (New Moon to New Moon)	29 days 13 hours
Diameter	3,476 km (2,160 miles)
Surface temperature	100°C (212°F) at noon -150°C (-240°F) at night
Force of gravity at surface	16 per cent of that on Earth
Atmosphere	none

LUNAR ECLIPSES

● At Full Moon, the Sun, Earth and Moon are sometimes in a straight line so that the Earth casts a shadow on the Moon. This is called a lunar eclipse. The Moon does not normally disappear from sight completely, but looks dark red or orange because some

Date		Time of mid-eclipse (EST)	Type	Duration of total eclipse (mins)
1992	Jun 15	9:58 a.m.	partial	
	Dec 9	11:45 p.m.	total	74
1993	Jun 4	1:01 p.m.	total	96
	Nov 29	6:27 p.m.	total	56
1994	May 25	3:32 p.m.	partial	
1995	Apr 15	12:16 p.m.	partial	

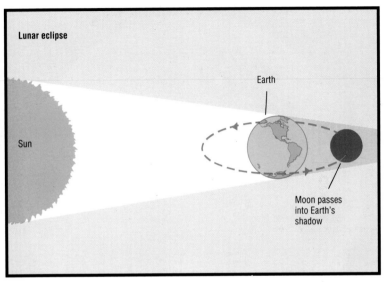

Lunar eclipse

Earth

Sun

Moon passes
into Earth's
shadow

There is an eclipse of the Moon when the Sun,
Moon and Earth line up and the Moon passes
through the Earth's shadow. (Not drawn to scale.)

sunlight still reaches it after
scattering from the Earth's
atmosphere. When a lunar
eclipse is taking place, it can be
seen from anywhere on Earth,
where the Moon has risen.

THE LUNAR LANDSCAPE

● **Maria**: flat areas of dark
volcanic rock. The large darker
areas visible to the naked eye
are maria. Before the telescope
was invented, it was thought
that these areas were seas of
water. Maria is the Latin word
meaning 'seas'. It is the plural
form of 'mare' (pronounced
maah-ray). Some dark areas
were given other watery names

in Latin, such *Oceanus
Procellarum* (Ocean of Storms)
and *Palus Nebularum* (Marsh of
Mists).

● **Craters**: circular hollows
with a raised wall around, and
sometimes one or more peaks
in the middle. Most lunar
craters were created by the
impact of rocks falling from
space long ago, soon after the
Moon was formed. Some
craters were flooded or
completely covered when
molten rock flowed out from
the Moon's interior to make the
maria.

● **Mountains**: these occur
particularly round the edges of
the large maria.

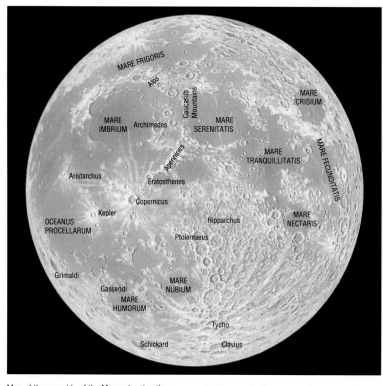

Map of the nearside of the Moon showing the 'seas' (labeled in capital letters), and the craters.

DID YOU KNOW ?

The highest mountains on the Moon, the lunar Apennines, reach over 7,500 m (24,600 ft).

● One of the most interesting times to observe a particular lunar feature is when it is on the boundary between the dark and bright parts of the Moon, the region called the 'terminator'. The shadows cast by any raised features are very long at this time, helping to show them up most clearly.

This photograph of Edwin Aldrin was taken by Neil Armstrong, near the *Apollo 11*'s lunar module.

MOON LANDINGS

● In the 1960s, both the USA and the former USSR developed spacecraft to investigate the Moon. The rivalry between the two great powers was seen by some as a 'space race', but it is doubtful whether the USSR ever intended to put a Russian cosmonaut on the Moon. It is difficult to know because the Soviet work was shrouded in secrecy. Major successes were announced, but the world rarely heard anything about plans or failures. It was the Americans who announced in 1961 that they were preparing to land a man on the Moon before 1970.

● The first steps in lunar exploration were taken by the former USSR. They launched a series of spacecraft towards the Moon, first called *Lunik*, then *Luna*. In 1959, *Lunik 3* took the first photographs of the far side of the Moon and, in 1966, *Luna 9* soft-landed on the Moon, proving that it has a firm surface. Some scientists had feared that the Moon was covered by a thick layer of soft dust, which would have made landing very difficult, if not impossible.

● Throughout the 1960s, the Americans prepared for a

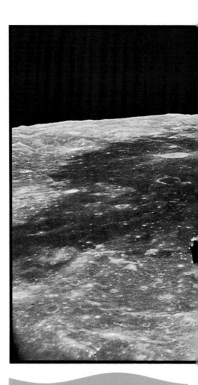

▶ **DID YOU KNOW ?**

On Earth an astronaut in his spacesuit weighs 135 kg (300 lb); on the Moon he weighs only 23 kg (50 lb).

landing. The final stage was the *Apollo* project. In 1968, *Apollo 8* orbited the Moon with three astronauts on board. On 21 July 1969 the historic moment came.

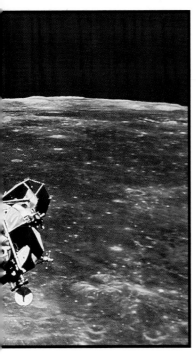

Apollo 11 photograph showing the lunar module. In the distance you can see the Earth rising over the Moon's horizon.

The Apollo missions had been a thrilling episode in the exploration of space - but an enormously expensive one too. The Apollo program alone was estimated to have cost $25 billion. The cost of the Soviet space program between 1958 and 1973 has been estimated at $45 billion.

● The *Apollo* crews left several 'moonquake' detectors on the surface of the Moon. Scientists have used the measurements of rock tremors, transmitted back to Earth, to work out the structure inside the Moon. There seems to be a rocky crust about 30 miles thick. Under that there is a rocky mantle about 560 miles thick. The very center of the Moon may be partly molten, perhaps with a small iron-rich, solid core.

Neil Armstrong stepped out of the lunar module of *Apollo 11* and put a foot on the Moon. His words were beamed all over the world, 'That's one small step for man, one giant leap for mankind.' Later he said that he had made a mistake. He intended to say 'a man', which would have made his meaning much clearer.

● In 1972 America sent a last manned flight to the Moon.

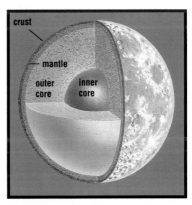

Cross-section of the Moon.

▶ ASTEROIDS

● In a band between the orbits of Mars and Jupiter, thousands of small rocks called asteroids (or minor planets) travel around the Sun. Over four thousand have been identified individually, so far.

● The largest asteroid, Ceres, is about 913 km (567 miles) in diameter. It orbits the Sun every 4·6 years at an average distance of 416 million km (259 million miles).

The minor planets, called asteroids, orbit mainly between Mars and Jupiter.

The largest asteroids

Name	Diameter (miles)
Ceres	566
Pallas	324
Vesta	310
Hygeia	266
Davida	209
Interamnia	206
Europa	193
Eunomia	168
Sylvia	168
Psyche	163

● Some asteroids do not orbit the Sun within the asteroid belt, but have long elliptical orbits that bring them very close to the Earth and the Sun. One of these is **Icarus**, which is only 1·4 km (0·9 miles) across. It travels to within 29 million km (18 million miles) of the Sun, much closer than Mercury, and glows red hot. In 1968 Icarus passed within 6·4 million km (4 million miles) of the Earth: a close encounter in astronomical terms.

● Small asteroids sometimes get even closer to the Earth. For example, one called **1989 FC** came within 690,000 km (400,000 miles) of the Earth in August 1989. It measured only 300 m (1,000 ft) across. If a rock of this size hit the Earth, it could explode with the force of about 2,000 megatons (2 billion tons) of TNT – 100,000 times greater than the atomic bomb that devastated Hiroshima in 1945.

▶ **METEORS AND METEORITES**

● Enormous numbers of particles called meteoroids orbit the Sun. They range in size from microscopic particles to dust to larger pieces of rock. They can only be seen when they enter the Earth's atmosphere, at speeds up to 43 mp/s, and burn up producing brief streaks of light known as meteors or shooting stars.

● Some meteors can be seen on most dark nights, although they may be quite faint. The chances of seeing one get greater after midnight.

● On certain nights more meteors than normal can be seen. When the Earth passes through a stream of meteoroids we experience a meteor shower. The shower meteors seem to come from one point in the sky, called the radiant. In fact this is a trick of the eye: the meteors enter the atmosphere traveling only in the same direction, and not from the same point. The meteoroid streams are trails of particles scattered along the path taken by a comet.

● On 13 November 1833, the Earth passed through a swarm of particles travelling in the orbit of comet Tempel-Tuttle. More than 30,000 meteors were seen in an hour. Amazing displays like this are very rare, though. They happen when the meteoroids are clumped together instead of being spread all round the comet's orbit.

Time exposure photo of the Leonid meteor shower. The Leonid shower occurs in November each year, when Earth enters an orbiting stream of debris from the comet Tempel-Tuttle.

● Meteor showers come round regularly on about the same dates each year, when the Earth in its orbit crosses through the meteoroid stream. The showers are named after the constellations in which their radiants lie. One of the best each year is the Perseids, which peaks on about 12 August.

● Most meteoroids are about the size of a grain of sand and are quickly burnt up when they enter the Earth's atmosphere. A larger object may burn particularly brightly and leave a trail that can be seen for several seconds. Meteors of this kind are called 'fireballs'. If they do not burn up completely, but fall to the ground, they are called meteorites.

Selected Meteor shower facts

Name	Peak date (and normal limits)	Maximum rate per hour	Associated comet
Quadrantids*	4 Jan (1-10 Jan)	60	Unknown
Lyrids	21 Apr (19-25 Apr)	10	Thatcher (1861 I)
Eta-Aquarids	5 May (24 Apr-20 May)	35	Halley
Perseids	12 Aug (23 Jul-20 Aug)	75	Swift-Tuttle (1862 II)
Orionids	22 Oct (16-27 Oct)	25	Halley
Leonids	17 Nov (15-20 Nov)	10	Tempel-Tuttle (1866 I)
Geminids	13 Dec (7-16 Dec)	75	Phaethon†

* This shower was named after the old constellation name Quadrans (the Quadrant), which is no longer used. The radiant lies in what is now Boötes, the Herdsman.
† Phaethon, discovered in 1983 and classified as an asteroid, is probably the nucleus of a 'dead' comet.

The maximum rates apply when the radiant is directly overhead, and the sky is very dark and moonless. At other times, the observed numbers are likely to be many fewer.

Meteorites are made of stone (aerotites), or metal (siderites), or both (siderolites). The siderites contain mainly the metals iron and nickel. This aerotite weighs 15 lb, and is approximately 1,3 billion years old.

● About 500 meteorites hit the Earth each year. Most of them are only the size of small stones. Fortunately, there has only ever been one report of someone being hit by a meteorite. Many of the larger meteorites came originally from the asteroid belt, but some are definitely made of Moon rock, and others seem to have come from Mars.

● Large meteorites that crashed onto the Earth long ago made craters, very much like the ones we can still see today on the Moon and other planets. Erosion by wind and water has long since wiped out most of them. The remains of about one thousand circular structures caused by meteorite impacts are known about. They range in size from 3 – 150 km (2 – 95 miles) in size, and are mostly found on the parts of the Earth with the oldest rocks.

● Meteor Crater in Arizona, USA (also known as Coon Butte or the Barringer Crater) is 1,265 m (4,150 ft) wide, 175 m (575 ft) deep, and surrounded by a wall 30 – 45 m (100 – 150 ft) high. Scientists estimate that the meteorite that made it about 50,000 years ago originally weighed 10,000 tons,

Aerial view of Meteor Crater, Arizona.

but most of it was destroyed in the impact. Scattered fragments of iron totaling 18 tons have been found in the area.

● In 1908, a fireball reported to be as bright as the Sun was seen falling towards the Earth over the remote Tunguska Valley in Siberia. It was seen from hundreds of miles away and a massive explosion shook the ground. However, the first expedition to the area of the explosion did not take place until 1927. It found that trees had been snapped off or knocked to the ground, and stripped of their branches over a region 30 – 40 km (20 – 25 miles) across, but there was no crater and no meteorite. The most likely explanation is that a comet nucleus exploded

The scene of devastation left by the Tunguska meteorite. People as far away as 750 km (466 miles) saw it in broad daylight.

about 5 miles above the ground. Fortunately, there was no loss of human life because the area was uninhabited.

● There are four parts of the world where glassy objects called **tektites** are found: Australia and south-east Asia, the Ivory Coast, Czechoslovakia and the states of Georgia and Texas in the USA. Tektites are thought to be formed from Earth rock melted in the blasts of meteorite impacts. Individual tektites can be as massive as 15 kg (33lb), but most are much smaller, like pebbles.

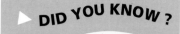

DID YOU KNOW?

Every day about 3,000 tons of dusty material from space falls onto the Earth.

● The tiny particles that fall to Earth from space are called micrometeorites, and are too small to burn up in the atmosphere. Although it sounds like a lot, the amount of material falling is very small indeed, compared with the total mass of the Earth. In a million years it will amount to less than one billionth the mass of the Earth.

► COMETS

● The nucleus of a comet is only a few miles across. It is made of frozen water, carbon dioxide, methane and ammonia, with dust and rocky material buried in it. Comets are often described jokingly as 'dirty snowballs'. As a comet approaches the Sun, the heat causes the ice to start evaporating. Gas is given off, forming a large sphere around the nucleus, called the coma. The coma may be up to a million miles across. Dust and gas are given off from jets on the side of the nucleus facing the Sun. They stream off into space, under the Sun's

Comet West was a bright comet, visible to the naked eye, seen in 1975. The two distinct tails can be seen. The dust tail shines by reflecting sunlight, while the straight charged-particle tail glows with its own bluish light.

Some well-known comets

Short-period comets

Name	Period (years)	
Biela	6.6	Broke in two about 1846. Last seen 1852.
Encke	3.3	Comet with shortest known period.
Halley	76	Recorded observations going back over 2,200 years.

Long-period comets

Name	Year	
Arend-Roland	1957	Dust in its orbit made it appear to have 'spike'.
Bennett	1970	Very bright with long tail.
Ikeya-Seki	1965	Sungrazer: passed through Sun's outer layers.
Morehouse	1908	Tails grew and broke off several times.
West	1976	Nucleus broke into four pieces.

influence, to form a tail, sometimes as much as a hundred million miles long. The tail has two parts. Electrically charged particles form a straight tail, swept along by the charged particles streaming outwards from the Sun. Particles of dust are gently pushed out by the pressure of sunlight itself to form another tail that is often broad and flat.

● Many comets are in elliptical orbits within the Solar System, that regularly bring them near the Sun, where they can be seen from the Earth every few years. These are called short-period comets. Each time they pass close to the Sun they lose some of their material so, over time, they break up and disappear completely.

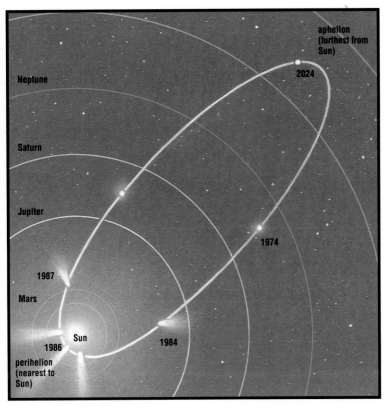

The orbit of Halley's comet around the Sun. At its most distant it passes beyond the orbit of Neptune.

● **Halley's** comet returns to the inner part of the Solar System every 76 years. Records show that it has been observed for at least 2,200 years. The last time it passed close to the Sun was in 1986. It is named after the English astronomer, Edmond Halley, who calculated its orbit and predicted that it would be seen in 1758, which turned out to be 16 years after his death. Its long elliptical orbit brings it within 56 million miles of the Sun at its closest approach and takes it out beyond the orbit of Neptune. When nearest the Sun it travels at about 3000 km/h (1,865 mph), but it slows down as it gets further away, down to 180 km/h (112 mph).

● In March 1986, the European space probe *Giotto* flew to within 375 miles of the nucleus of Halley's comet, and returned images of it.

Artist's impression of the *Giotto* spacecraft as it flies within 375 miles of the nucleus of Halley's Comet.

● The closest encounter the Earth has had with a comet was in 1770 when Lexell's comet came within 1,200,000 km (750,000 miles).

● Comets are thought to come from a region surrounding the Solar System at a distance of about one light year.

▶ **DID YOU KNOW ?**

About 20 comets are observed each year but most of them are very faint, and can only be seen with the aid of a telescope.

STARS

▶ **STARS**

● Compared with the size of the Solar System, the distances to even the nearest stars are enormous. The nearest star, Proxima Centauri, is just over 40,000,000,000,000 km (25,000,000,000,000 miles) away.

● Such distances are so great, they are often given in light years rather than miles . **A light year is the distance light travels in one Earth year**. The speed of light is 186,000 miles per second so in one year it goes 5,878,000,000,000 miles. This makes Proxima Centauri 4¼ light years away.

▶ **DID YOU KNOW ?**

It would take a car travelling 100 mph more than 29 million years to reach our nearest star.

The ten nearest stars

Star name	Constellation	Distance (light years)
Proxima Centauri	Centaurus	4·3
Alpha1 Centauri	Centaurus	4·3
Alpha 2 Centauri	Centaurus	4·3
Barnard's Star	Ophiuchus	6·0
Wolf 359	Leo	7·8
Lalande 21185	Ursa Major	8·2
UV Ceti A	Cetus	8·4
UV Ceti B	Cetus	8·4
Sirius A	Canis Major	8·6
Sirius B	Canis Major	8·6

The letters 'A' and 'B' after a star name refer to the different members of a double star system.

Light we see from this star now left it 4¼ years ago.

● It takes light from the Sun about 8¼ minutes to reach the Earth.

STAR BRIGHTNESS

● Some stars in the night sky are obviously brighter than others and a telescope reveals many stars invisible to the naked eye. Astronomers describe the brightness of a star as its **magnitude**. The brightest stars seen with the naked eye are mostly about magnitude 1 (though a few are even brighter) and the faintest are magnitude 6. The larger the number of the magnitude scale, the fainter the star.

● A magnitude 1 star is 2·512 times brighter than a

The bright star you can see here is Vega, the fifth brightest star in the sky. It is in the constellation of Lyra.

magnitude 2 star and a magnitude 2 star is 2·512 times brighter than a magnitude 3 star, and so on. The number 2·512 multiplied by itself 5 times makes 100, so a magnitude 1 star is exactly 100 times brighter than a magnitude 6 star.   Objects brighter than magnitude 1 may be 0, or a negative number on the magnitude scale. The brightest star Sirius, has a magnitude of -1·5; the Sun has a magnitude of -26·8, and a full Moon about -12.

● With typical binoculars you can see stars of about magnitude 9, and a powerful telescope may be able to reach down to magnitude 27.

● The apparent brightness of a star as seen from Earth depends on how far away the star is as well as the amount of light it actually emits. Stars that look faint might be very bright, but a long way away. The true relative brightness of a star is given by its **absolute magnitude**. This is defined as

The ten brightest stars				
Star name	Constellation	Magnitude apparent	absolute	Distance (light years)
Sirius	Canis Major	-1·5	1·4	8·6
Canopus	Carina	-0·7	-8·5	1170
Alpha Centauri	Centaurus	-0·3	4·4	4·3
Arcturus	Boötes	0·0	-0·2	36
Vega	Lyra	0·0	0·5	26
Rigel	Orion	0·1	-7·1	910
Capella	Auriga	0·1	0·3	42
Procyon	Canis Minor	0·4	2·6	11
Achernar	Eridanus	0·5	-1·6	85
Betelgeuse	Orion	0·5	-5·6	390

the **apparent magnitude** it would have if it were at a standard distance of 32·616 light years away from us.

● The Sun's absolute magnitude is 4·8. The most luminous stars are around -8 and the feeblest about 16.

STAR COLORS

● The color of a star depends on its temperature. Very hot stars with surface temperatures over 18,000°F glow white. Cool stars, at less than 5,500°F are red. The Sun, somewhere between with a surface temperature of 11,000°F looks yellow.

● Astronomers group stars into 'spectral types' according to the appearance of their spectra. They use a spectrograph to measure the spectrum of a star. The spectrograph breaks up the star's light and spreads it into a spectrum. A star's surface temperature is the main influence on the general appearance of its spectrum. This means that that the different spectral types roughly correspond to different star colors. Each group is known by a letter of the alphabet, but not in alphabetical order. When astronomers first tried (around a hundred years ago) to group stars in order of temperature, they used the letters of the alphabet, but made mistakes and got the types in the wrong

The main spectral types		
Type	Colour	Typical surface temperature (°F)
O	bluish white	45,000
B	bluish white	27,000
A	white	18,000
F	yellowish white	12,600
G	yellow	9,900
K	orange	7,200
M	red	5,400

	STELLAR TEMPERATURE						
	72,000°F	54,000°F	18,000°F	13,500°F	10,800°F	8,850°F	6,300°F	4,350°F

The Hertzsprung-Russell diagram shows the types of star and how bright they are.

(Diagram: absolute magnitude plotted against spectral class, with stars labeled)

SUPERGIANTS – Ia
Naos, Saiph, Rigel, Aludra, Deneb, Wezen, Betelgeuse
Mimosa, Adhara, SUPERGIANTS – Ib, Canopus, Mirfak, Polaris, Enif, Antares
Spica, Achernar, BRIGHT GIANTS – II, Suhail, Almach, Gacrux
Regulus, Algol, Pollux, Capella, Dubhe, Aldebaran, Kocab, Mira
ABSOLUTE MAGNITUDE, Vega, Castor, Sirius A, GIANTS – III, Arcturus, SUBGIANTS – IV
Fomalhaut, Altair, Procyon A
MAIN SEQUENCE, Rigil Kent, Sun, Centauri B
Eridani
61 Cygni A
61 Cygni B
Sirius B, Kapteyn's star, Lalande 21185
WHITE DWARFS, Procyon B, Bernard's star, Ross 128
Van Maanen's star, Proxima Centauri

Absolute magnitude scale: -8, -6, -4, -2, 0, 2, 4, 6, 8, 10, 12, 14, 16

SPECTRAL CLASS: O, B, A, F, G, K, M (0 5 for each)

order. The same letters are still used for the groups, but in order of decreasing temperature, from hottest to coolest.

● If you choose a sample of stars, say all the stars in a star cluster or all the stars within a certain distance of the Sun, it is interesting to plot a graph of true brightness (absolute magnitude) against spectral type. Each star is represented by its own point on the graph. Such a graph is called a Hertzsprung-Russell diagram after the two astronomers who first thought of it.

● It turns out that the points are not scattered randomly. Most are in quite a narrow band sweeping diagonally from the top left to the bottom right. This band is called the main sequence and consists of the ordinary stars. The Sun is a main-sequence star.

● Above the main sequence are points corresponding to **red giant** and **supergiant** stars and below it the **white dwarfs**. These are stars that used to be on the main sequence but have undergone changes as they got older.

● All stars appear as points of light in a telescope. Apart from the Sun, none is near enough for its disc to be seen directly. However, by studying spectra and measuring the absolute magnitudes of stars, it is possible to work out that some are hundreds of times larger than the Sun, while others are as small as a planet.

Stars in the sky do not look very different in size, but some are hundreds of times larger than the Sun, and others are only the size of Earth.

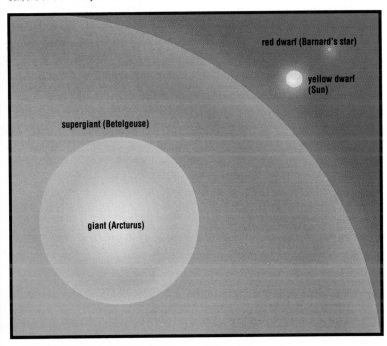

red dwarf (Barnard's star)

yellow dwarf (Sun)

supergiant (Betelgeuse)

giant (Arcturus)

LIFE AND DEATH OF A
STAR LIKE THE SUN

Formed in the clouds of gas and dust in space.

The collapsed central part of the star, a **white dwarf**, is all that remains. About the size of the Earth, it contains almost the mass of the Sun, so that a teaspoonful would weigh over a ton. The white dwarf gradually cools and fades.

Starts to run out of hydrogen in the core. The core shrinks and the outer layers expand to make the star a **red giant**, a hundred times larger than the present size of the Sun.

Star's outer layers thrown off to form a shell of glowing gas, known as a **planetary nebula**.

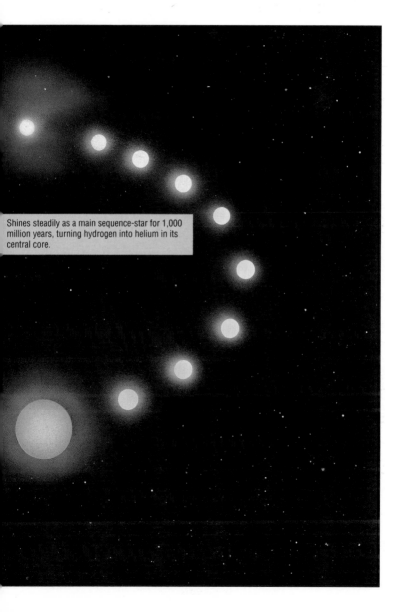

Shines steadily as a main sequence-star for 1,000 million years, turning hydrogen into helium in its central core.

SUPERNOVAE

● A star several times more massive than the Sun ends its life dramatically as a supernova. When its energy source suddenly runs out, the star effectively destroys itself in a massive explosion. Suddenly, the star increases in brightness to shine like a million Suns, then gradually fades over a period of months. The gas that has been blown off disperses into space over time. All that is left of the original star is a tiny neutron star a few miles across.

● In 1987 there was a supernova bright enough to be seen with the naked eye. It exploded in the Large Magellanic Cloud, a nearby galaxy about 170,000 light years away and visible from the southern hemisphere.

Before

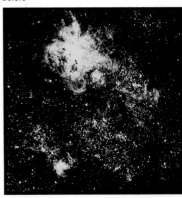

On the right of the lower image is a bright star, supernova 1987A. The upper photo was taken in 1969, and the lower one in 1987 two days after the explosion of a blue supergiant star. Supernova 1987A was the brightest seen since 1607. The Large Magellanic Cloud is top center.

After

▶ **DID YOU KNOW ?**

The Crab Nebula is the remains of a supernova that exploded in AD 1054.

NEUTRON STARS AND PULSARS

● When stars several times more massive than the Sun reach the end of their lives, they explode as supernovae. All that remains after a supernova is the collapsed core, a small, dense star about 25 km (15 miles) across. In these strange stars, the

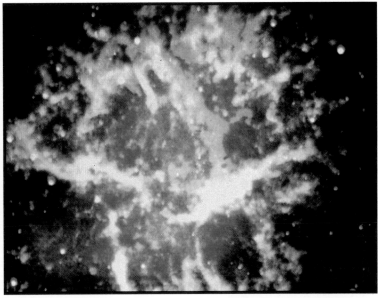

Color-enhanced photo of the Crab Nebula, a supernova remnant about 6,000 light years away in the constellation of Taurus. The nebula is the remnant of a catastrophic stellar explosion observed by Chinese astronomers in 1054 AD.

electrons and protons that normally make up ordinary atoms are crushed together to make neutrons, so they are called **neutron stars**. A spoonful of their material would weigh over a billion tons on Earth.

● Scientists predicted that neutron stars would spin very quickly – about once a second or even faster. They also expected that pulses of radio waves might be detected as beams from the spinning stars swept into view; rather like the effect of a beam from a lighthouse.

● In 1967, the first **pulsar** was

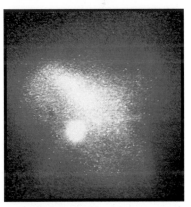

X-ray image of the Crab Nebula. The bright spot at lower left is the Crab pulsar. It is the remnant of the star which exploded in 1054 AD.

discovered flashing a burst of radio waves towards the Earth every 1·34 seconds. The radio astronomers called it a 'pulsar' because it was a pulsating source of radio waves. More pulsars were discovered and soon there was little doubt that they are spinning neutron stars. Several hundred pulsars are now known, including one in the Crab Nebula, where a supernova exploded in AD 1054, and another in the remains of a supernova ten thousand years old in the constellation Vela. Some of them flash invisible light, X-rays and gamma-rays, as well as radio waves.

QUASARS

● The word quasar is short for 'quasi-stellar radio objects' which means star-like emitters of radio waves.

False-color radio image of quasar 3C 273, which is 2,100 million light years away, and our nearest quasar.

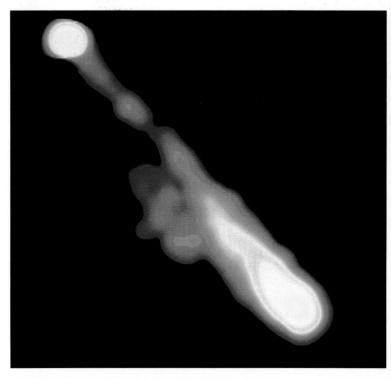

● In the 1960's astronomers were amazed when they discovered that some objects in the sky, which looked like tiny stars, gave out more energy than a whole galaxy. They called them **quasars**.

● Some quasars are as far away as we can see into the universe using the most powerful telescopes – about 15,000 million light years away.

● The brightest quasar, called 3C 273, is about 2,000 million light years away and is in the center of a large galaxy. Many quasars have been found in the center of galaxies.

> **DID YOU KNOW ?**
>
> Some scientists think that the quasar's energy might come from a black hole in the center of it.

BLACK HOLES

● If the core of a star left over after a supernova contains more than three times the Sun's mass, so much material gets concentrated in so small a volume of space that a star-sized 'black hole' is created. Around a black hole, the force of gravity is so great that nothing can escape – including light.

● Although black holes cannot be seen directly, it should be possible to tell if one is present in a double star system. Material sucked in by the black hole would be heated up so much that X-rays would be emitted. Though X-ray binary (double) stars are known, no black hole has been identified for certain. The X-ray binaries might have white dwarfs or neutron stars in them, rather than black holes.

DOUBLE STARS

● About half the 'stars' in the sky are actually two or more stars orbiting around each other, like the planets orbit the Sun. Double stars are also called **binary systems**. Many multiple systems of three or more stars also exist. The stars in such a system can be similar, or quite different in size and type.

● In some **close binaries**, the stars are almost touching, with material streaming from one to another, or a cloud of gas surrounding both of them. The closer the stars, the faster they complete an orbit. An X-ray binary star is known with an orbital period of only 11 minutes 25 seconds. Stars as close as this cannot be seen separately in any telescope. Astronomers have to work out what is happening by studying the spectra and other kinds of evidence put together from observations.

● At the other end of the scale, there are pairs of stars that are far enough apart to be seen as doubles. They may take hundreds or thousands of years to orbit around each other.

● Some binary stars are lined up from our point of view on Earth, so that we see one star alternately pass behind and in front of the other. Systems like this are called **eclipsing binaries**. The most famous eclipsing binary is the star Algol in the constellation Perseus. Its magnitude dips from 2·2 to a minimum of 3·5 every 2·87 days as an eclipse occurs.

Eclipsing binaries are stars with orbits that mean they pass in front of each other.

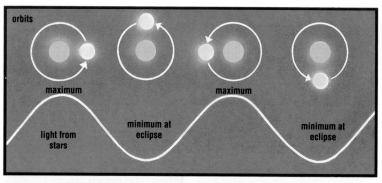

Optical doubles are stars that appear close together in the sky, but are in fact a long way apart.

● Sometimes two stars appear to be close together in the sky when they are really separated by a vast distance, and not connected with each other at all. Pairs like this are called **optical doubles**.

VARIABLE STARS

● Some stars vary in brightness from time to time, either more or less regularly or in a totally unpredictable way.

● **Cepheid variables** are a type that vary in a regular way over a few days. The name comes from the star Delta Cephei, which varies between magnitudes 3·6 and 4·3 over a period of 5·4 days. The change takes place as the unstable star pulses in and out. Many Cepheids are known, with periods ranging between 3 and 50 days.

● **Long-period variables** are red giant stars that change over time-scales between 100 and 1,000 days. Both the brightness change and the period over which it takes place can be different from one cycle to the next. Some show very dramatic variations. One of the most well-known is Mira in the constellation Cetus, which varies between magnitudes 2 and 10 over about 332 days. It goes from being as bright as the Pole Star, to hardly detectable in binoculars.

NOVAE

● Novae are stars that suddenly increase in brightness by about ten magnitudes, then fade again over a period of months. The name 'nova' comes from the Latin *nova stella*

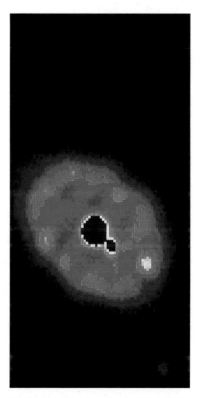

The remains of Nova Herculis, which exploded in 1934. It is surrounded by gas thrown out by the explosion.

meaning 'new star'. Novae are not uncommon, but most never get bright enough to be seen by the naked eye. Perhaps one every three years or so, might be that bright. The brightest ever recorded was in the constellation Aquila in 1918. It changed from being an 11th magnitude star to being almost the brightest star in the sky, in only six days.

● Novae are binary stars in which one of the pair is a white dwarf. The outburst is due to a huge nuclear explosion on the surface of the white dwarf, triggered by a build-up of material streaming from the companion star.

▶ GALAXIES

● Look up at the night sky and you might imagine that stars are scattered all over the Universe. Stars are actually found only in large groups called galaxies. Galaxies cover a huge size range, from just a few thousand light years to hundreds of thousands of light years across. The smallest contain about a hundred thousand stars and the largest thousands of billions.

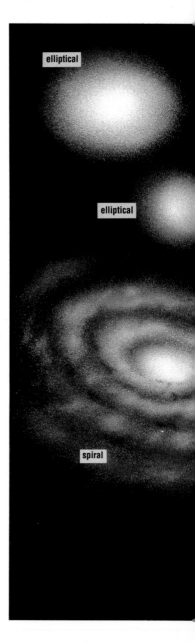

Astronomers classify galaxies by appearance and shape. Elliptical galaxies can vary in shape, from almost spherical, to quite elongated.

barred spiral

irregular

● All the stars you see with the naked eye in the night sky, belong to our own Galaxy (written with a capital letter to distinguish it from other galaxies). The Galaxy is roughly disc-shaped with a bulge at the middle and some stars scattered in a sphere all around, called the halo. The Sun is situated in the disc, about two thirds of the way out from the center. From inside it, we see the disc of stars as the **Milky Way**, a faint band of light circling the sky. Our Galaxy is sometimes called the Milky Way Galaxy.

● Surveys of the sky made with powerful telescopes show that millions of galaxies populate the Universe for as far as can be seen. The Universe is expanding so the galaxies are all speeding apart. The further away a galaxy is from us, the faster it is receding. The most remote galaxies visible are going at almost the speed of light.

● There are three main types of galaxy: **elliptical**, **spiral** and **irregular**. Most normal galaxies are elliptical and about a quarter of all galaxies are irregular. Our own Galaxy is a spiral. In some spiral galaxies, the spiral arms seem to start at the ends of a bar extending from the central bulge of stars. These galaxies are called barred spirals.

● Galaxies rotate, but not like solid objects. Each star is in its own individual orbit around the center, like the planets are in orbit around the Sun. The Sun takes 220 million years to make a complete trip around the center of the Galaxy.

● Some galaxies have an exceptionally powerful source of energy concentrated at their centers. Such galaxies are described by astronomers as 'active'. They include **radio galaxies**, which emit strong radio signals from two huge clouds either side of them, and quasars, many of which are also powerful radio sources. Quasars are the most luminous known objects in the Universe. Their energy appears to come from a region at their center no bigger than the solar system. Almost certainly, this energy source is a supermassive black hole with matter falling into it.

● The nearest galaxy to our own is the **Large Magellanic Cloud**. It is an irregular galaxy about 30,000 light years across and 170,000 light years away. The **Small Magellanic Cloud** is about 20,000 light years across and 200,000 light years distant.

> **DID YOU KNOW ?**

The most massive known galaxy, called M87, is in the Virgo Cluster of galaxies at a distance of about 60 million light years.

The Large Magellanic Cloud, a companion galaxy of our own Milky Way. Many pink nebulae can be seen. The largest one, far left, is the Tarantula Nebula.

CLUSTERS OF GALAXIES

● Galaxies are not scattered randomly through the Universe. Clusters of galaxies are held together by the pull of gravity each member exerts on the others. Our Galaxy belongs to a small cluster known as the **Local Group**. About twenty-five member galaxies have so far been discovered. In contrast, the **Virgo Cluster** contains several thousand galaxies. The **Coma Cluster** at a distance of 350 million light years is another example of a

The Coma Cluster in the constellation Coma Berenices.

rich cluster with more than a thousand bright galaxies.

● The Andromeda galaxy, a spiral like our own Galaxy but twice as big, is the largest member of the Local Group. Two hundred thousand light years across and about 2·3 million light years away, it is

The Andromeda Galaxy is a giant spiral galaxy which dominates the Local Group, of which our own Milky Way galaxy is a member.

the most distant object visible to the naked eye.

● Clusters of galaxies are themselves linked together in 'superclusters'. The Local Group lies near the edge of the Local Supercluster, which is more than a hundred million light years across and centered on the Virgo Cluster.

THE MILKY WAY GALAXY

● On a dark, moonless night you will be able to see a faint white band of light across the sky. This is called the Milky Way. With binoculars or a telescope you can tell that it is

A x-ray image of The Milky Way. The different colors show the varying amount of x-ray emission from the galaxy. Yellow is the most intense, and blue is least intense.

made up of numerous individual stars. There are about two hundred billion stars altogether in the Galaxy. Two spiral arms seem to wind outwards from the Galaxy's central bulge. Viewed from distant space, our Galaxy would look like a typical spiral.

● A side view of the Galaxy shows a thin disc with a bulge at the center. The disc is about 1,000 light years thick and the bulge about 20,000 light years across. There is a scattering of stars and star clusters throughout a 'halo' that surrounds the bulge, and is about 100,000 light years across. Clouds of dust prevent us from seeing the centre of the galaxy directly. Scientists have

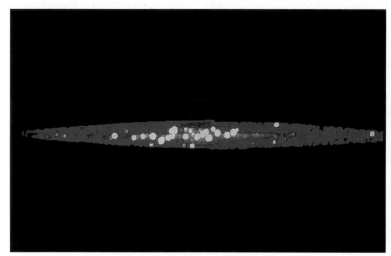

The Milky Way is shaped like a thin disc with a bulge in the center. Stars, dust and gas fan out from the center bulge to form a spiral.

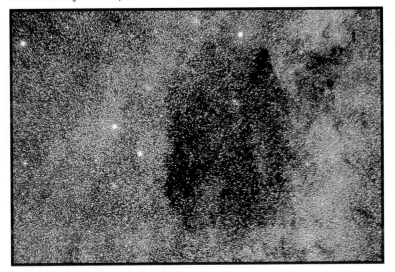

The Coalsack nebula is a large dark cloud of dust in the Milky Way, in the Southern Cross. It blocks the light of the stars behind it.

The Pleiades is situated in the constellation of Taurus. It is a young cluster on a galactic time scale, being only about 50 million years old.

used infrared and radio telescopes to probe this region.

● Most of the stars in the Galaxy lie in the central bulge and the spiral arms. Younger stars and regions where new stars are forming, are concentrated in the spiral arms. Clusters of new stars are often surrounded by glowing clouds of hydrogen gas called emission nebulae. The ultraviolet light given out by the hot stars causes the hydrogen to glow with a pinkish light.

● The loose clusters of young stars in the disc of the Galaxy are called open clusters. They may contain between a few hundred and a few thousand stars. Thousands of such clusters are known: some of them, such as the Pleiades and the Hyades in Taurus and Praesepe in Cancer, are visible to the naked eye.

● The **Pleiades** is an open cluster of stars about 400 light years away. The popular name for this cluster is 'The Seven Sisters', but most people can only make out six individual stars with the naked eye. There are between 300 and 500 stars in the cluster in all. The haze surrounding the stars is a reflection nebula of gas and dust. The bluish light is reflected starlight: the nebula does not glow with any light of its own.

The Orion Nebula consists of several nebulae. Some are bright clouds of gas and dust where stars are in the process of being born. It is located 1600 light years away in the constellation of Orion, and spreads out over 15 light years of space. Radiation from the new stars in the Nebula light up its gas, causing it to glow red.

● The Galaxy's 'halo' contains most of its oldest stars, many of which are concentrated in tightly-packed globular clusters. More than a hundred globular clusters have been identified by astronomers. Each contains hundreds of thousands, or even millions, of stars.

● A substantial amount of the material in the Milky Way is in the form of dust and gas clouds, but the material between the stars is spread out very thinly indeed. The best vacuum that can be produced on Earth contains more material than is typically found between stars.

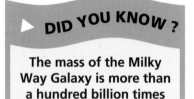

▶ DID YOU KNOW ?

The mass of the Milky Way Galaxy is more than a hundred billion times that of the Sun.

Omega Centauri, situated in the constellation of Centaurus, is the brightest globular cluster in the sky. It is 620 light years in diameter and 16,500 light years away. This photograph was taken in ultraviolet light using a telescope on shuttle *Columbia*, in December 1990.

SPACE TRAVEL

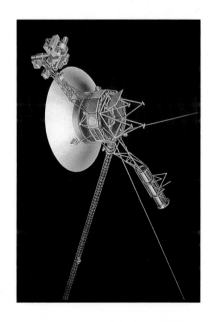

▶ SATELLITES

Some important 'firsts'

1957
● The first artificial satellite, *Sputnik I*, was launched by the USSR on 4 October. It was about the size of a soccerball and traveled around the Earth every 96 minutes, at a height of between 220 km (135 miles) and 930 km (580 miles). A small radio transmitter powered by a chemical battery was on board. It sent out a constant 'bleep-bleep' signal, which was picked up around the world, but within a few days the battery was dead.

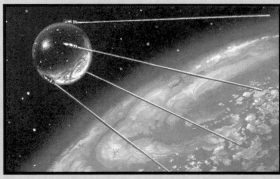

Sputnik 1

● In November, *Sputnik II* was launched with a dog named Laika on board. She was the first living creature sent into space, but she died when the oxygen supply ran out.

1958
● In February the USA launched its first satellite, *Explorer I*. It measured 15 cm (6 in) in diameter and carried both a radio transmitter and a Geiger counter for measuring radiation from space. The findings from this satellite led to the discovery of two ring-shaped regions around the Earth where electrically charged particles (electrons and protons) are trapped by

the Earth's magnetic field. These regions are called the Van Allen belts after the scientist who discovered them.

● In March the USA launched the experimental satellite *Vanguard I*. It tested a new power system using solar cells that absorbed sunlight and converted it into electrical power. Today, most satellites are powered by means of thousands of solar cells, that together can generate thousands of watts.

1962

● *Telstar*, the first communications satellite, was launched by the USA on 10 July. It cost three million dollars, and provided Europe with the first live television pictures from America. *Telstar* only operated for one year.

Artist's impression of a
Telstar 4
communications
satellite

● Today, communications satellites link all parts of the world. We can see news almost as it happens, even in remote places. We have instant telephone communications almost everywhere, and computers exchange information with each other via satellites.

● The network of satellites providing worldwide telecommunications is called **Intelsat** (International Telecommunications Satellite Organization). The Intelsat satellites are placed in orbits 35,900 km (22,300 miles) above the equator. In this position the satellite travels round the Earth in the same time that it takes the Earth to rotate, so the satellite appears to hover at about the same point in the sky the whole time. Such an orbit is described as 'geostationary'.

● One of the important uses of satellites is keeping a close watch on the planet Earth. About half of all the satellites launched by the USA and former USSR have been for military purposes. Some have huge dishes for eavesdropping

Intelsat 5 in geostationary orbit over the Atlantic.

on radio communications. Others track the movements of tanks, ships and submarines.

● Satellites are also used for monitoring the weather, and keeping a check on environmental changes: charting the courses of oil slicks, measuring the rate at which forests are destroyed, and the loss of vegetation in drought-stricken areas. With the help of instruments to detect a wide range of different chemicals in the atmosphere, satellites can monitor pollution and changes to the ozone layer.

● Images from satellites are used to make detailed maps of the Earth's surface, which help geologists find oil and rare

Satellite receiver-transmitter dishes at Goonhilly Earth Station in southwest England. These dishes receive and transmit information to a satellite serving *Inmarsat* (the International Maritime Satellite Organization), which provides telecommunication links for ships at sea, and aircraft.

minerals. Such maps are also used by engineers when designing, for example, irrigation projects or major new roads in a remote part of the world.

False-color satellite image of a severe storm, in the Bering Sea, photographed by the *Nimbus 5* weather satellite.

▶ SPACE PROBES

SPACE PROBES TO VENUS

● Venus was the first planet to be reached by a space probe, only four years after the first successful satellite launch by the USA.

1962
● In December, *Mariner 2* (USA) flew within 35,400 km (22,000 miles) of the surface of Venus. It was the first successful space probe, and transmitted back to Earth information about the planet's temperature. At 480°C (900°F) it was much hotter than scientists had expected. *Mariner 2* also recorded details about the atmosphere of Venus, measured its rotation period and discovered that it has hardly any magnetic field.

1965
● *Venera 3* (USSR) reached Venus. It was designed to make a parachute landing on the planet's surface but was crushed during its descent by the enormous atmospheric pressure.

1970
● *Venera 7* (USSR) was the first space probe to land on Venus. Although it landed successfully and transmitted data from the planet's surface, it was put out of action within an hour by the very high temperature.

1972
● *Venera 8* landed on surface and transmitted data back to earth.

1975
● *Veneras 9* and *10* landed on 21 and 25 October, respectively. Each transmitted only one picture before ceasing to work, but these were the first pictures from Venus's surface.

1978
● Two *Pioneer Venus* (USA) spacecraft reached Venus. The first was an orbiter carrying radar, which mapped much of the surface. The other was a 'multiprobe' consisting of five small landers to measure atmospheric data; they were

not intended to function after reaching the surface.

1982

● *Venera 13* (USSR) transmitted the first color pictures from the surface of Venus.

1990

● In August, the *Magellan* spacecraft (USA) arrived in orbit around Venus and began providing scientists with detailed radar images which would eventually cover the whole planet.

● The *Magellan* radar images reveal detail ten times finer than the best previously obtained. They record objects 120 m (400 ft) across. The best pictures from *Veneras 15* and *16* did not resolve anything less than 1,600 m (1 mile) across.

● *Magellan* has also been used to study variations in gravity over Venus's surface, which will help scientists understand more about the planet's interior.

Artist's impression of NASA's *Magellan* spacecraft surveying the planet Venus. *Magellan* was launched in 1989.

SPACE PROBES TO MARS

1962
● The first attempt to send a space probe to Mars, by the USSR, failed a few weeks after launch.

1964
● *Mariner 3*, the first attempt by the USA to send a probe to Mars, went out of control and was lost, but *Mariner 4*, launched on 28 November successfully reached Mars the following year.

1965
● On 14 July, *Mariner 4* passed within 9,800 km (6,120 miles) of the surface of Mars. It had traveled a total distance of 670 million km (418 million miles) even though Mars comes within 56 million km (35 million miles) of Earth. This was done to minimize the amount of fuel needed and make best use of natural motion under gravity. It took pictures over a few days, and revealed the presence of craters on Mars.

1971
● In November, *Mariner 9* (USA) reached Mars and went into orbit around the planet. At its closest, it was 1,370 km (850 miles) above the surface. The first pictures were puzzling because they showed so little detail. It was soon realized that the planet was engulfed in a dust storm. A few months later the dust settled and *Mariner 9* then transmitted more than 7,000 pictures over a period of about a year.

1971-73
● The USSR made at least six attempts to land a space probe on the surface of Mars. All failed.

1976
● *Viking*s 1 and 2 (USA) became the first space probes to land successfully on the surface of Mars. Each consisted of an orbiter and a lander. After entering orbit around the planet, the two parts of the probe separated and the lander descended through the martian atmosphere. At the top of the atmosphere it was traveling at

about 16,000 km/h (10,000 mph) but, with the aid of a parachute and rockets, it gently landed at about 10 km/h (6 mph).

● The *Viking* landers sent back images, analyzed gases in the atmosphere and chemicals in scoops of soil; measured wind speeds and atmospheric pressure. Meanwhile, the orbiters mapped the Martian surface.

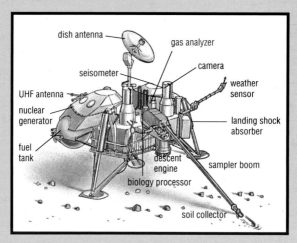

Viking Lander

SPACE PROBE TO MERCURY

1974

● *Mariner 10* (USA) was the first space probe to visit Mercury. It made its first flyby of the planet in 29 March, and sent back over 600 images of the surface. *Mariner 10* remained over Mercury for 17 hours and then continued round the Sun to meet up with Mercury again six months later, and for a third time in March 1975. Soon after the third encounter, *Mariner 10*'s equipment failed, though it is probably still orbiting the Sun. *Mariner 10* was the first space probe to fly by more than one planet: it was able to visit Venus on the way to Mercury.

SPACE PROBES TO JUPITER AND BEYOND

● Sending space probes to Jupiter and the other outer planets posed many problems. They had to be reliable for many years since it takes almost ten years to reach Uranus and a further three to get to Neptune. The spacecraft could not use solar cells as a source of power because not enough sunlight reaches the outer Solar System. Instead, nuclear-powered generators had to be provided, using uranium or plutonium as fuel.

● There was also the worry that a spacecraft might be destroyed as it passed through the asteroid belt. Even a small rock could cause devastating damage in a collision. Fortunately, all spacecraft so far have passed through the asteroid belt without any such encounters.

1973	● On 3 December, *Pioneer 10* (USA) flew past Jupiter 132,000 km (82,000 miles) above the cloud tops. It transmitted images, measured the temperature of the planet's atmosphere, and mapped Jupiter's extensive magnetic field.
1974	● On 2 December, *Pioneer 11* flew past Jupiter and then continued to Saturn, where it arrived in 1979, and successfully returned images of the planet and rings.
1977	● *Voyagers* 1 and 2 (USA) were launched. Both were extremely successful. *Voyager 1* explored Jupiter (1979) and Saturn (1980), and their satellites. *Voyager 2* was able to continue to Uranus (1986) and Neptune (1989) after visiting Jupiter (1979) and Saturn (1981).
1989	● In October, *Galileo* was launched from the Space Shuttle to reach Jupiter in 1995. In 1991 it returned the first close-up image of an asteroid, Gaspra.

Voyager 2

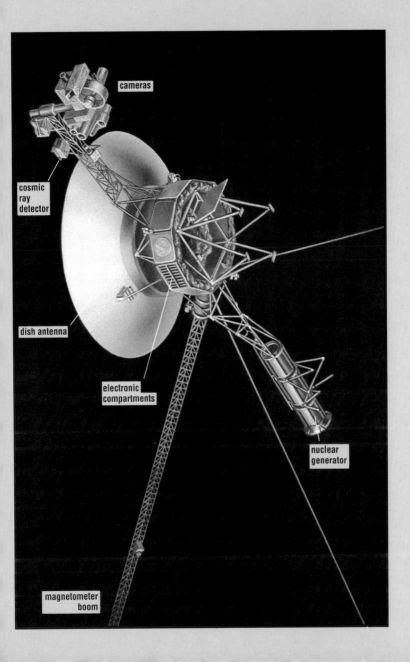

cameras

cosmic
ray
detector

dish antenna

electronic
compartments

nuclear
generator

magnetometer
boom

HUMANS IN SPACE

1960
● In August, a Soviet spacecraft carried two dogs, Belka and Strelka, into space, and returned them safely to Earth. They were the first living creatures to be sent into space and brought back.

1961
● On 12 April, the first person to travel into space, Yuri Gagarin, was launched in the Soviet spacecraft *Vostok 1*. He returned safely to Earth 1 hour 48 minutes later.

● On 5 May, the first American went into space. Alan Shepard made a straight up and down journey (without going into orbit), that lasted 15 minutes.

1962
● John Glenn became the first American astronaut to orbit the Earth.

1963
● Valentina Tereshkova aboard *Vostok 6* (USSR) made the first space flight by a woman.

1965
● In March, the USSR launched *Voskhod 2* and cosmonaut Alexei Leonov left the spacecraft to become the first person to walk in space.

● On 3 June, Ed White made the first space walk by an American. This was part of the Gemini project in preparation for sending men to the Moon.

1967
● On 27 January, during a practice countdown for *Apollo 1* (USA) the cockpit caught fire and the astronauts Ed White, Virgil Grissom and Roger Chaffee died.

1968
● On 21 December, *Apollo 8* (USA) blasted off with astronauts Frank Borman, James Lovell and William Anders on board. They made the first manned orbit of the Moon and, at their closest, passed only 110 km (69 miles) above the Moon's surface. They splashed down safely in the ocean on 27 December.

1969

● On 16 July *Apollo 11* (USA) was launched to make the first manned landing on the Moon. The astronauts on board were Neil Armstrong, Edwin 'Buzz' Aldrin and Michael Collins. The spacecraft went into orbit 110 km (70 miles) above the Moon's surface. On 20 July, Aldrin and Armstrong entered the Lunar Module, which then separated and descended to the surface, leaving Collins in orbit in the Command Module. At 10:56 pm EDT on 21 July, Neil Armstrong took his first step onto the Moon. He was followed by Buzz Aldrin. They spent three hours on the surface setting up experiments and taking soil samples.

1970-72

● Further manned *Apollo* landings on the Moon followed. The last was *Apollo 17*, which returned to Earth in December 1972.

1983

● Sally Ride, on board the Space Shuttle Columbia, became the first US woman in space.

Apollo 11 is launched, beginning the first lunar landing mission.

● Getting people to the Moon required new, very powerful rockets. The *Saturn V* rockets used for this purpose, were the world's most powerful. They were 111 m (366 ft) tall, almost the height of the Great Pyramid in Egypt. They weighed about 3,000 tons at lift-off but most of the weight was fuel that was used up before the spacecraft left the Earth's atmosphere. The spacecraft that reached the Moon weighed only about 14 tons.

● The first stage of the rocket ignited at blast-off. The thrust generated was equivalent to the full power of the engines of 50 Boeing 747 Jumbo Jets. It lifted the rocket to a height of about 66 km (41 miles) then fell away, having burnt out.

● The second stage then ignited and lifted the rocket further. It too fell away when burnt out.

● The third stage accelerated the spacecraft to 28,800 km/h (18,000 mph), fast enough to escape from the Earth's gravity. This speed was reached about 11 minutes after leaving the launch pad. The journey to the Moon took about three days.

● The *Apollo* spacecraft consisted of three parts: a Command Module, a Service Module and a Lunar Module. The Command and Service Modules remained in orbit round the Moon during the landing. The Service Module was jettisoned shortly before the craft re-entered the Earth's atmosphere.

SPACE STATIONS

● Space stations are large orbiting satellites that stay in space for years at a time. Crews of astronauts take turns at living and working in them, perhaps for a few months or even as long as a year.

● The USA has had only one space station, *Skylab*, launched in 1973. The Soviet Union put a series of seven *Salyut* space stations in orbit between 1971 and 1982 and launched the *Mir* space station in 1986. Soviet cosmonauts have spent long periods working in space. One cosmonaut, Yuri Romanenko, has lived for a total of 430 days in space over three separate trips.

Valentina Tereshkova, the first woman in space, in front of the *Vostok 6* spacecraft in which she made 48 revolutions of the Earth, in 1963.

Yuri Gagarin in the cabin of *Vostok 1*, in which he became the first person to orbit the Earth, in 1961.

1971   ● *Salyut 1* (USSR), the first space station, was launched in April. It was manned for 23 days by three cosmonauts, who were killed in an accident on the return flight to Earth in *Soyuz II.*

1973   ● *Skylab* was launched by the USA on 14 May. It was 6 m (20 ft) long and 6·4 m (21 ft) wide. A crew of three astronauts manned *Skylab*. The first crew stayed for 28 days, the second for 59 days and the third for 84 days. They performed scientific experiments and carried out detailed observations of the Earth and Sun, taking more than 180,000 photographs of the Sun. They also studied the effects that living in space had on themselves. In the weightless conditions experienced in orbit, all muscles including the heart become weaker and start to waste away. The skeleton becomes very brittle. Some of these effects are reduced by doing exercises, but they still pose problems for really long journeys through space.

1974   ● *Salyut 3* (USSR), the first military space station, and *Salyut 4* were launched.

1975-82   ● *Salyut*s 5, 6 and 7 were launched. Soviet cosmonauts began staying in space stations for much longer, with crews remaining in *Salyut 6* for over six months and in *Salyut 7* for almost a year.

1986   ● *Mir* (USSR), a very large space station, was launched. It was designed to have extra modules added but the plans were delayed due to shortage of money and technical problems.

Because of the weightless conditions in space, astronauts have to be fully zipped into sleeping bags, and the bags fixed to the wall of the spacecraft, so they do not float away.

The US space station *Skylab*. The photograph was taken from the Command Module as the crew were leaving *Skylab* for the last time. In July 1989 *Skylab* entered the Earth's atmosphere and burnt up. A few fragments fell to the ground in Western Australia, fortunately in an uninhabited part and no one was hurt.

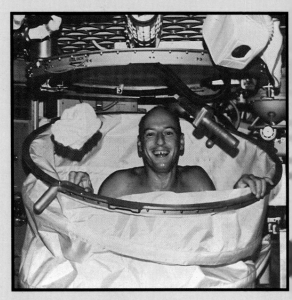

Having a shower is a problem without gravity; the water forms globules and floats into the air. Here astronaut Charles Conrad has just finished showering. When in use the shower curtain is pulled right up from floor to ceiling. Water is removed by vacuum.

THE SPACE SHUTTLE

1981	● The first Space Shuttle, *Columbia*, was launched by the USA with astronauts John Young and Roger Crippen aboard. They remained in space for 54 hours.
1984	● A Shuttle launch was stopped a few seconds before take-off by a computer detecting a fuel problem.
1986	● On 28 January, the Space Shuttle *Challenger* exploded 73 seconds after take-off, killing the crew of seven. The Space Shuttle program was stopped until a careful investigation of safety had taken place.
1988	● The first successful launch of the Space Shuttle took place following the 1986 tragedy.
	● The USSR launched their first unmanned reusable spacecraft, *Buran*.

● The Space Shuttle is a spacecraft that can be used many times, making it an economical form of space transport.

● The winged Shuttle orbiter looks like an aeroplane. At the front in the nose of the Shuttle is the flight deck where the pilot and co-pilot sit, and beneath them are living quarters with a kitchen, toilet and sleeping area. Behind the nose is the large body of the Shuttle; this is a storage area in which satellites can be transported. It is the cheapest way of launching a satellite into orbit, and the Shuttle can also be used to collect faulty satellites from space and return them to Earth for repair. Loads of 29 tons can be carried into orbit, and about 14 tons can be brought back.

● Launching the Shuttle into orbit requires two big booster rockets, which fall off about two minutes after launch, when their fuel supply is used up. The boosters have parachutes and fall into the sea where they are recovered and used again. The huge fuel tank drops off a few minutes later. It goes into orbit and eventually falls into the atmosphere where it breaks up. It is the only part of the Shuttle that cannot be reused.

● The Shuttle can carry up to eight astronauts into space where they can do experiments and even go outside the spacecraft to repair satellites already in orbit.

● The Shuttle returns to Earth like a glider. As it falls through the Earth's atmosphere it is slowed by resistance from the air. This produces temperatures of over 1,500°C (2,700°F) at the outer hull of the Shuttle. At the end of its descent, the Shuttle lands gently on a long runway.

Orbiter Discovery on the launch pad with external fuel tank and two booster rockets. It carried a crew of six astronauts and three satellites.

Astronauts can move around outside the Shuttle wearing a rocket-powered back-pack called a Manned Maneuvering Unit. In 1984 Bruce McCandless (seen here) was the first astronaut to move around independently in space. He flew 100 m (330 ft) away from the Space Shuttle.

STAR GAZING

▶ TELESCOPES

● We see the Universe as it looks in visible light, the kind of radiation our eyes can detect. Optical telescopes collect more light than our eyes can pick up on their own, and so help astronomers see fainter objects. The bigger a telescope the more light it can collect and the fainter the things it can see. Photographs and electronic light detectors also help to get the best results from a telescope.

● But today's astronomers are not only collecting visible light. Stars and galaxies, and the material between them, give out radiation across the whole of the electromagnetic spectrum: radio, and micro waves, and infrared with wavelengths longer than visible light; ultraviolet light, X-rays and gamma-rays with shorter wavelengths. There are telescopes to collect all these different kinds of radiation.

● Only visible light and radio waves pass easily through the Earth's atmosphere. Though some places are much more suitable than others, optical and radio telescopes can work almost anywhere on the Earth. Radio waves even pass through clouds.

● On high mountain sites, where the air is very clear and dry, it is possible to have infrared and micro-wave telescopes. However, infrared observations are best done from telescopes orbiting the Earth in space, well above the atmosphere. Ultraviolet light, X-rays and gamma-rays cannot get throught the atmosphere at all and have to be studied from orbiting observatories.

● Even for optical telescopes, the atmosphere creates a problem. Star images constantly twinkle and look blurred rather than as sharp points, and dust and clouds can make them look dimmer. In April 1990 the Hubble Space Telescope was put in orbit by the Space Shuttle. It is a reflecting telescope with a mirror 2·4 m (94·5 in) across. After it was launched, astronomers discovered that a mistake in shaping the mirror meant that images could not be focused as sharply as they should have been. Nevertheless, the Space Telescope has already produced some remarkable pictures, much better than anything taken from the ground, and it should be possible to correct the fault on a future space mission.

OPTICAL TELESCOPES

● There are two main kinds of optical telescopes: reflectors and refractors. **Reflectors** use a curved mirror to collect light. **Refractors** have a main lens, called an 'objective'. The size of the objective is called the telescope's 'aperture'. The size of the aperture is the chief factor determining the faintest objects the telescope can see. The aperture of the human eye is about 8 mm (0·3 inches).

● The largest refracting telescope in the world was completed in 1897 and is at the Yerkes Observatory near Chicago in the USA. Its objective is 1·01 m (40 in) across. Larger telescopes are all reflectors because it is impossible to make suitable giant lenses and hold them in place in a telescope.

● The largest reflecting telescope with a single mirror is in the Caucasus region between the Black Sea and the Caspian Sea, at the Special Astrophysical Observatory

A reflecting telescope (right), and a refracting telescope.

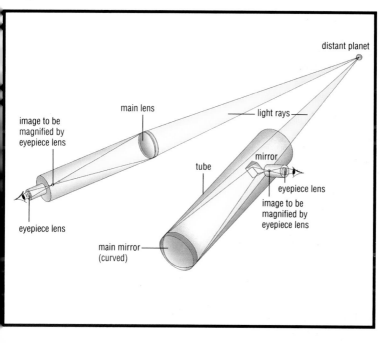

distant planet

main lens

light rays

image to be magnified by eyepiece lens

tube

mirror

eyepiece lens

image to be magnified by eyepiece lens

eyepiece lens

main mirror (curved)

near Zelenchukskaya. The mirror's diameter is almost 20 feet.

● The Multiple Mirror Telescope on Mount Hopkins in Arizona has six individual mirrors, each 1·8 m (72 in) across, that can work together to give the light-gathering power of a 4·5-metre (176-in) telescope.

The Yerkes Observatory near Chicago.

● The Keck Telescope being constructed in Hawaii has a mirror 33 feet across made of 36 individual hexagonal segments fitted together.

● Work began in 1988 on a set of four 8-m (300-in) telescopes at a site in Chile. It will be possible to link the telescopes to observe together, giving them power equivalent to a single mirror 16 m (52 ft) in diameter.

● Special solar telescopes are

Clear aperture m in		Name	Location	Date operational
6·00	236	Bolshoi Teleskop Azimutalnyi	Special Astrophysical Obs- ervatory, Zelenchuckskaya	1976
5·08	200	Hale Telescope	Mount Palomar, California	1948
4·20	160	William Herschel Telescope	La Palma, Canary Islands	1987
4·00	158	Inter-American Observatory	Cerro Tololo, Chile	1976
3·89	153	Anglo-Australian	New South Wales, Australia	1975
3·81	150	Mayall Reflector	Kitt Peak, Arizona	1973
3·80	150	UK Infrared Telescope	Mauna Kea, Hawaii	1979
3·60	142	Canada-France- Hawaii Telescope	Mauna Kea, Hawaii	1979
3·57	141	European Southern Observatory	La Silla, Chile	1976
3·50	138	German-Spanish Astronomy Centre	Calar Alto, Spain	1983

The largest single-mirror reflecting telescopes

The 200 inch Hale telescope in its dome at Mount Palomar Observatory.

The Cerro Tololo Inter-American Observatory sited in the foothills of the Andes.

used for studying the Sun. The largest is the McMath Solar Telescope at Kitt Peak National Observatory in Arizona. It has a 1·5-m (60-in) mirror, mounted on a tower, which directs sunlight along a tilted tunnel, much of which is below ground. It produces an image of the Sun 75 cm (30 in) in diameter.

RADIO TELESCOPES

● A radio telescope consists of a radio antenna that collects the weak radio signals from space and feeds them into an amplifier and radio receiver. Many radio telescopes have very large reflecting dishes as their antennas. Sets ('arrays') of such dishes can be linked together to make the telescope more powerful.

● The largest radio dish in the world is at the Arecibo Observatory in Puerto Rico. A dish 305 m (1,000 ft) across has been built into a natural depression in some hills. It cannot be moved, and can only look at objects passing over it in the sky as the Earth rotates on its axis.

A view from beneath the wire mesh forming the 1,000 ft diameter radio telescope at Arecibo, Puerto Rico. It is the largest radio dish in the world.

● The largest fully-steerable radio dish is at Effelsberg in Germany and is 100 m (328 ft) in diameter

The largest single-dish radio telescopes	
1000 ft	Arecibo, Puerto Rico
328 ft	Effelsberg, Germany
250 ft	Jodrell Bank, UK
210 ft	Parkes, New South Wales, Australia
150 ft	Algonquin Observatory, Ontario, Canada

● The largest array of radio dishes is the Very Large Array in New Mexico, USA. It consists of 27 movable dishes, each 25 m (82 ft) across, arranged in a Y-shape with each arm 21 km (13 miles) long.

▶ **CONSTELLATIONS**

● Looking at the night sky, it is possible to see patterns made by the brightest stars, for example lines, squares and crosses. Over 2,000 years ago, the Greeks identified certain groups and patterns of stars with various animals and characters from their myths. In about AD 140, the Greek astronomer Ptolemy listed over 40 constellations, including Orion (The Hunter) and Ursa Major (the Great Bear).

● Since Ptolemy's time, new constellations have been added and astronomers today use 88, which cover the whole of the sky between them. Each constellation is a particular area of sky, not just the pattern of bright stars within it.

● Stars that look close to each other in the sky are not necessarily near in space. The patterns we see from Earth are for the most part chance alignments, with some stars much further away than others.

● Many of the brighter stars have individual names: Sirius, Capella and Algol, for example. From the 17th century, astronomers have also used star names that combine the constellation name with a letter of the Greek alphabet. Sirius is also Alpha Canis Majoris (Alpha in Canis Major, the Great Dog) and Algol is Beta Persei (Beta in Perseus).

● With the help of a simple star chart, you can start to pick out the patterns of bright stars. Some patterns that stand out particularly well make good 'signposts' to the other stars.

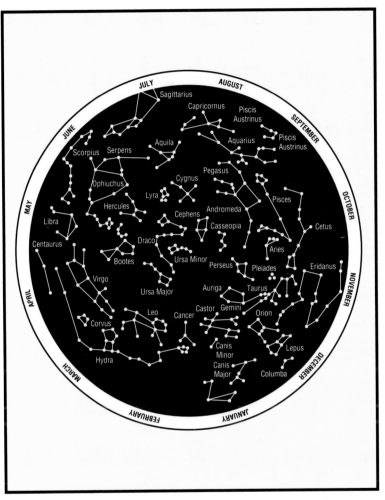

The major stars and constellations of the Northern Hemisphere. To use the map, face south and turn the page so the month you are in, is at the bottom.

For instance, following the line made by the three stars of Orion's belt leads you to Aldebaran, the brightest star in Taurus, the Bull, and the star cluster of the Hyades, which appears roughly V-shaped.

Following the line through further, you come to another cluster, the Pleiades, also called Seven Sisters.

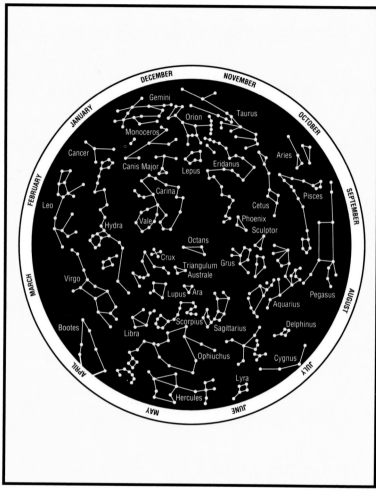

The major stars and constellation of the Southern Hemisphere. To use this map, in the Southern Hemisphere, face north, and turn the map so the current month is at the bottom.

● Orion and Taurus are prominent constellations visible from the northern hemisphere in winter evenings. The constellations you can see at any particular time gradually change from day to day, so what you can see in the night sky in summer and winter is completely different.

BC

747
● Babylonians began records of eclipses. They noticed that solar eclipses occurred at certain regular intervals which were multiples of a period of 17 years 11 ⅓ days. This list was used 900 years later by the famous Greek astronomer Ptolemy (see 140 AD).

720
● Chinese records of eclipses began.

This clay tablet from 7th century BC, is Assyrian. It is an astronomical forecast based on observations of the Sun.

585
● Thales of Miletus, a Greek astronomer, correctly predicted a solar eclipse, possibly using the Babylonian observations (see 747 BC). His prediction caused a great deal of interest especially as the eclipse took place during a battle between the Medes and Lydians, and the two armies immediately stopped fighting and established peace between their nations.

547
● Death of Anaximander (born 611 BC). He was a pupil of Thales (see 585 BC), and made one of the first maps of the surface of the Earth. He thought that we live on a flat disc at one end of a drum-shape, which is floating in space at the center of the universe.

400
● About this time Greek thinkers suggested that the Earth is a sphere, because a ship slowly disappears below the horizon.

347
● Death of Plato (born 427 BC). He was a Greek philosopher, whose ideas ruled astronomy for

The Tower of the Winds in Athens, built in 1 BC, is an eight-sided building, with each side facing the direction of one of the winds. It also contained sundials and a waterclock. The Greeks believed that through science and observation they could understand the universe.

2000 years (until the work of Kepler, see 1609). He believed that the Earth is at the center of the universe, and the heavens are a perfect place in which only perfect circular movements are possible.

356 ● Death of Eudoxus (born 409). He was a pupil of Plato and worked out a model for the way the planets move round the Earth. This model became the basis of astronomy for 2000 years. He suggested that each planet moves as if it is attached to a sphere, and each sphere is fixed to other spheres, rather like the layers of an onion. At the center of all the spheres is the Earth.

230 ● Death of Aristarchus (born about 310). He stated that the Earth and all the planets move round the Sun. This view was not accepted, and the Earth remained at the center of the solar system until Copernicus (see AD 1543). He made the first scientific attempt to measure the distance of the Sun and Moon from the Earth. He estimated that the Sun is 18 times further away from us than the Moon. In fact it is 390 times more distant.

120 ● Death of Hipparchus (born about 190). He set up an observatory on the island of Rhodes. His detailed observations enabled him to work out the length of a solar year very accurately. His

measurements also gave the average length of a lunar month as 29 days 12 hours 44 minutes and 2 ½ seconds, which is less than a second out compared to present calculations!

● Hipparchus made accurate estimates of the size and distance of the Moon from Earth. He also compiled a catalogue of 850 stars.

AD

100

● Ptolemy (Claudius Ptolomaeus of Alexandria) was born. He wrote one of the most important books in the history of science, it became known as the *Almagest* meaning 'the greatest'. In it he used mathematics to account for all the movements of the planets. It was accurate enough to explain all that can be seen with the naked eye, and his theory remained unchallenged for about 1500 years.

829

● An observatory was built at Baghdad. The greatest of all Arabic astronomers Al-Battani (died 929) worked there making very accurate observations of the Sun and Moon.

1054

● The Chinese recorded a supernova which remained visible for 22 months. Today it forms the Crab Nebula.

1066

● A large comet was observed, and thought by King Harold of England to be a bad omen,

The scene in the Bayeux Tapestry where King Harold is told of the comet.

warning of defeat at the hands of William the
Conqueror. It is shown on the Bayeux Tapestry.
This comet later became known as Halley's
comet (see 1705).

1449

● Ulugh Beg (born 1394), a Mongol astronomer,
was assassinated by his son. He had built an
observatory in Samarkand and very accurately
listed the positions of stars.

The observatory of
Ulugh Beg. This 131 ft
marble arc was used to
measure the positions
of stars.

1543

● Shortly before his death, Nicolas Copernicus
(1473–1543) published a book explaining his
theory that the Earth and other planets move
round the Sun. He had completed this book ten
years earlier, but was nervous to publish an idea
that was the opposite of what everyone then
believed (that the Earth did not move and was
at the center of the universe). Surprisingly his
book did not cause a great stir at the time but
later astronomers, such as Galileo (see 1633),
who made discoveries which supported
Copernicus's ideas, were persecuted.

1572

● Tycho Brahe (1546–1601) observed a new star, he called it a nova (today it would be called a supernova). He estimated its distance and found that it was beyond the moon and planets and in the far off region of the stars. This presented a major problem because at this time people believed that the heavens were unchanging and nothing new could appear. Any sudden bright lights in the sky like a new star or comet could only be produced close to the Earth and not beyond the Moon.

1577

● Tycho Brahe observed a comet and showed that it was further away than the Moon (see 1572).

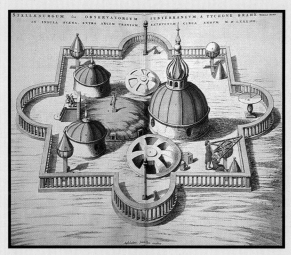

Tycho Brahe's observatory built on an island between Sweden and Denmark. Brahe's observations, made with the naked eye, were extremely accurate.

1580

● King Frederick II of Denmark built an observatory for Tycho Brahe on the island of Hven between Denmark and Sweden. Brahe recorded the most accurate observations ever made with the naked eye. (The first astronomical telescope was invented 7 years after Brahe's death). His careful observations of the positions of the planets would eventually be used by his assistant Kepler to give the first accurate picture of the elliptical orbits of the planets (see 1609).

1609

● Publication of the *Astronomia Nova* (New Astronomy) by Johann Kepler (1571–1630). In his book he claimed that the planets move round the Sun in elliptical orbits.

● Galileo (1564–1642) built his own telescope based on a recent Dutch invention and used it to look at the sky. It had only previously been used for looking at things on Earth. He saw things that no one had ever been able to see before: moons of Jupiter; craters on our Moon; individual stars in the Milky Way, not just a band of white light; and the phases of Venus as it moved round the Sun reflecting its rays.

1633

● Galileo, old and with failing eyesight, was put on trial by the Roman Catholic Inquisition. To avoid torture and death Galileo declared that the Earth does not really move round the Sun. He was condemned to be a prisoner in his house for the rest of his life.

1656

● Christian Huygens (1629–1695) observed a ring around Saturn. The ring was first seen by Galileo but his telescope made it appear like two 'horns', one either side of the planet (see 1675).

1668

● Isaac Newton (1642–1727) invented the first reflecting telescope, which still forms the basic design of many of today's astronomical telescopes.

Isaac Newton's reflecting telescope.

1675

● The Greenwich Observatory was founded by King Charles II.

● Olaus Roemer (1644–1710) first showed that light has a definite speed, and measured it.

● Giovanni Cassini (1625–1712) observed two rings round Saturn. (See 1656 and 1980).

1687

● Publication of the greatest scientific book, the *Principia*, by Isaac Newton. His book gave a mathematical explanation of the structure and workings of the whole universe. His laws of universal gravitation made it possible to explain the structure of the Solar System.

1705

● Edmond Halley calculated that a bright comet he saw in 1682 had appeared in 1531 and 1607. He correctly predicted that it would return in 1758. It has become known as Halley's comet and is seen about every 75 ½ years. It was most recently seen in 1986.

An 18th century astronomical observatory, built by Sawai Jai Singh, in Jaipur.

1781   ● William Herschel (1738–1822) discovered the
                planet Uranus. At first he thought it was a
                comet. This was the first planet discovered with
                a telescope. This discovery was rewarded by a
                large pension from the king, George III. William
                and his sister Caroline (1850–1848) were then
                able to become full-time astronomers, no longer
                relying on their musical talents to earn money.
                They spent their lives devoted to careful
                observations of the night skies.

1783   ● William Herschel published his first
                catalogue of double stars, most of them had
                never been noticed before.

1786   ● The first volume of William Herschel's
                Catalogue of nebulae was published. It
                contained 1000 nebulae never recorded before.
                Three years later another 1000 were added and a
                further 500 were listed in 1802. These catalogues
                formed the basis of Dreyer's *New General
                Catalogue* of 1888 which is still used by
                astronomers today.

1789   ● William Herschel's new forty-foot reflecting
                telescope was completed. This giant telescope
                was a technological marvel but weighed a ton,
                and was very difficult to handle. Herschel did
                use it on 28 August 1789 and discovered a sixth
                satellite of Saturn, but generally he continued to
                use his more manageable twenty-foot telescope.

1801   ● The first asteroid, Ceres, was discovered by
                Giuseppe Piazze (1746–1826).

1821   ● Friedrich Bessel (1784–1846) began very
                accurate measurements of the positions of
                50,000 stars. The work was completed twelve
                years later.

1838   ● The first measurement of the distance of a
                star, other than the Sun, was published by
                Friedrich Bessel. His careful work showed that
                the star known as 61 Cygni is 10·3 light years

away. Today with much better instruments astronomers calculate a distance of just over 11 light years.

1843

● The existence of an undiscovered planet beyond Uranus was predicted by John Couch Adams (1819–1892). The orbit of Uranus did not quite fit Newton's Laws, and Adams calculated that the orbit was disturbed by a planet further out in the Solar System. Urbain Le Verrier (1811–1877) working independently in Paris also calculated the position of this unknown planet.

1846

● The planet Neptune was discovered by Johann Galle using the calculations of John Couch Adams and Urbain le Verrier.

1866

● Giovanni Virginio Schiaparelli (1835–1910) established that the orbits of certain comets and certain meteor showers are identical.

1868

● Angelo Secchi (1818–1878) completed a study of the spectra of about 4,000 stars. He placed all stars in one of four groups depending on whether they appeared blue, yellow, red or dark red. This formed the basis of the classification of stars used by astronomers today. The color of a star shows how old it is.

1885

● Charles Pritchard (1808–1895) published the first catalogue of stars giving accurate measurements of their brightness.

1905

● Percival Lowell (1855–1916) predicted that there is a planet beyond Neptune (see 1930).

● Albert Einstein's papers on the special theory of relativity were published. Among other things he stated that nothing can travel faster than the speed of light, and he also produced the equation $E=mc^2$, relating energy (E) to mass (m) and the speed of light (c).

1906

● An enormous mysterious explosion occurred near Tunguska in Siberia. Millions of trees were destroyed, but no remains of a meteorite were

The edge of the
Tunguska crater.

found. It has been suggested that it was caused by a comet or a small black hole.

1916   ● Einstein's general theory of relativity was published. One of its predictions was that a light ray passing close to a massive object like a star will be affected by its gravitational attraction, and be curved (see 1919).

1918   ● A 2·5 m (100 in) reflecting telescope was set up at the Mount Wilson Observatory, USA. It was the world's largest telescope until 1948. It could no longer be used after 1987 because of air pollution.

1919   ● Arthur Eddington (1882–1944) provided the first experimental proof that Einstein's general theory of relativity was correct (see 1916). During the solar eclipse of 29 May he showed that light rays passing close to the Sun were bent by the amount predicted by Einstein.

1929   ● Proof that the universe is expanding was given by Edwin Powell Hubble (1889–1953). Using observations made with the 100 inch reflecting telescope on Mount Wilson (see 1918) he showed that the more distant a galaxy is, the faster it is moving away from Earth.

1930   ● On 18 February the planet Pluto was discovered by Clyde William Tombaugh (see 1905).

1931
● Radio waves from outer space were discovered by radio engineer Karl Jansky (1905–1950). He was using a simple antenna to find out what was causing interference in radio telephone links. In 1933 he showed that the radio waves he had detected were actually coming from the Milky Way. This was the beginning of radio astronomy.

1933
● A new theory explaining how stars produce so much energy was proposed by Hans Bethe (1906–), and independently by Carl von Weizsacker (1912–). They suggested that it is a process of nuclear fusion changing hydrogen into helium, which produces the energy output of stars. This is the theory which scientists still accept.

1939
● The existence of black holes was predicted theoretically by J. Robert Oppenheimer (1904–1967).

1948
● Completion of the 200 inch Hale reflecting telescope at Palomar, California, USA. It was the best optical telescope in the world for about 40 years. There is a picture on page 120.

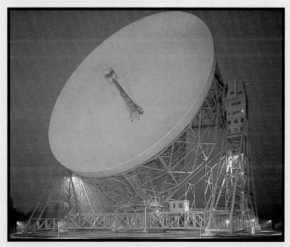

Night-time view of the 76 m (250 ft) diameter dish of the Lovell radio telescope at Jodrell Bank, Cheshire England.

● Hermann Bondi (1919–) and Fred Hoyle (1915–) suggested the steady-state theory of the universe in which matter is constantly created and forcing the universe to expand. Scientists now reject this theory and have adopted the Big Bang theory instead (see 1965).

1955   ● The radio telescope at Jodrell Bank, UK, was completed. It has a dish 76·2 m (250 ft) across.

1957   ● *Sputnik 1*, the first man-made satellite, was launched by the Soviet Union. Later in the year a satellite containing a dog was sent into orbit.

1958   ● The first American satellite was launched.

1959   ● The Russian satellite *Lunik 2* took the first photographs of the far side of the Moon.

1961   ● 12 April Yuri Gagarin (1934–1968) of the USSR became the first person in space. He orbited the Earth in *Vostok 1* for 1·8 hours.

● 5 May Alan Shepard, the first US astronaut in space, traveled round the Earth for 15 minutes.

1963   ● 16 June the first woman in space, Soviet cosmonaut Valentina Tereshkova Nildayeve, orbited the Earth 48 times in 78 hours.

● The first quasar (quasi-stellar radio source) was discovered.

1965   ● Soviet cosmonaut Leonov took the first spacewalk. It lasted 20 minutes.

● A background whisper of radio waves from throughout the universe is detected. Astronomers believe that this is the remains of the Big Bang which happened at the formation of the universe.

1966   ● 3 February Soviet spaceprobe *Luna 9* made the first soft landing on the Moon.

1967   ● Three US astronauts died in a fire on their Apollo spacecraft during a ground test on the launch pad.

● A Soviet cosmonaut died as his spacecraft *Soyuz 1* crashed to Earth on re-entry.

● Jocelyn Bell detected regular pulses of radio waves coming from outer space. Could they be signals from a distant life form? They were soon shown to be coming from spinning neutron stars and were named pulsars.

1969
● 20 July US astronaut Neil Armstrong became the first man to step onto the Moon.

1971
● US spacecraft *Mariner 9* went into orbit round Mars and sent back photographs. Two weeks later a Soviet spaceprobe began orbiting Mars.

1972
● The last manned landing on the Moon. Three US astronauts spent 44 hours on the Moon and collected over (220 lb) of rock samples.

1973
● The first US spacestation *Skylab* spent 27 days in space. The three man crew carried out scientific and medical experiments. Later in the year the second *Skylab* stayed in space for 59 days, and it was followed by a third which remained there for 84 days.

1975
● The Russian spaceprobes *Venera 9* and *Venera 10* sent back pictures from the surface of Mars.

1977
● *Voyager 1* and *Voyager 2* were launched on their journeys to the outer planets of the solar system.

1981
● US launched the first reusable spacecraft, *Columbia*. This was the beginning of the space shuttle program.

1982
● On space shuttle *Columbia's* fifth mission it carried a satellite into space.

1983
● The first American woman in space, Sally Ride, was a crew member of the second US space shuttle, *Challenger*.

1986
● Six astronauts and a woman school teacher were killed on board the space shuttle

Challenger when it exploded 73 seconds after launch.

● The first permanently occupied space station *Mir* (Peace), was launched by the Soviet Union.

● *Voyager 2* flew past Uranus.

1988   ● Soviet cosmonauts Vladimir Titov and Musa Manarov completed a record 365 days in space.

1989   ● *Voyager 2* flew past Neptune.

1989   ● *Magellan* spacecraft launched from space shuttle *Atlantis*. In August 1990 it reached the planet Venus. It was the first spaceprobe to be launched from a shuttle, and the first to be sent into space for more than a decade.

● A spacecraft to Jupiter was launched. Called *Galileo*, it is expected to reach Jupiter in 1995. *Galileo* consists of two parts: an entry probe and an orbiter. The entry probe will separate and descend into Jupiter's clouds sending back important information about the planet's atmosphere. It will operate for about 75 minutes.

Space Shuttle *Atlantis* blasts away from the Kennedy Space Center, Florida, 18 October, 1989. On board is the *Galileo* Jupiter probe.

1990   ● The Hubble Space Telescope, an orbiting observatory, was launched with a 15-year lifetime in space.

1991   ● Soviet cosmonauts were stranded in space during the break-up of the USSR.

1996   ● Expected launch of the spaceprobe *Cassini* to Saturn. It will reach the planet in 2002 and make a detailed study of the moon Titan.

absolute magnitude The brightness a star would appear to have if it were at a standard distance of 32.6 light years away.

aperture The diameter of the main lens or reflector of a telescope.

aphelion The point in the orbit of a planet (or other object) around the Sun, that is furthest from the Sun.

apogee The point in the orbit around the Earth of an artificial satellite or the Moon that is furthest from the Earth.

apparent magnitude The relative brightness of a star or other astronomical object as seen in the sky from Earth.

astronomical unit (AU or a.u.) The average distance between the Earth and the Sun. 1 AU = 149,597,870 km (about 92.96 million miles).

atmosphere The outer layers of gas surrounding a planet, satellite or star.

atom The smallest unit of a chemical element. Atoms are typically one ten millionth of a millimeter in size. An atom contains a central nucleus of protons and neutrons surrounded by a cloud of electrons.

axis The imaginary line through a planet, satellite, star or other object about which it rotates.

binary star A pair of stars in orbit around each other, held together by the gravitational attraction each has for the other.

black dwarf A star in the very final stage of its life when it is no longer radiating any energy.

black hole A region of space where the gravitational force is so strong that nothing, including light, can escape from it.

celestial sphere The sky pictured as the inside of a distant hollow sphere surrounding the Earth. The concept of the celestial sphere is used for describing the positions and movements of astronomical objects.

corona The outermost layers of the Sun, which become visible as a white halo during a total solar eclipse.

density A quantity that measures how compact a body is, that is how much mass is contained in a unit of volume. Density is measured in grams per cubic centimeter (g/cm^3) or kilograms per cubic meter (kg/m^3). The density of water is about 1 g/cm^3.

Doppler shift A change in the wavelength of sound waves or electromagnetic waves (such as light, radio etc.) due to the source of waves being in relative motion towards or away from the observer. The spectra of stars and galaxies moving away from us show a redshift.

dwarf star A star that is of normal size for its mass, or smaller, and is not a 'giant' or 'supergiant'.

ecliptic The plane of the Earth's orbit around the Sun. The Earth's orbital movement makes it appear as if the Sun travels around the sky once a year. The Sun's annual path through the stars is also known as the ecliptic.

electromagnetic radiation A form of energy that travels through a vacuum at a speed of 186,000 mp/s. It consists of linked electric and magnetic fields and its character varies with wavelength. From the longest to the shortest wavelengths, the main types are radio, infrared, visible, ultraviolet, X-rays and gamma-rays.

electron A small elementary particle that carries a negative electric charge.

escape velocity The minimum velocity with which an object needs to travel in order to overcome the pull of gravity of another object at a particular point in space, and so travel off into space independently. The velocity of escape from the surface of the Earth is about 7 mp/s.

galactic center The central region of the Milky Way Galaxy. It lies in the direction of the constellation Sagittarius and cannot be seen in visible light because of dense dust clouds. It is marked by a strong radio source called Sagittarius A and is thought to contain a massive concentration of stars, possibly even a black hole.

gravity An attractive force that acts between all material objects.

light year The distance traveled by light in one Earth year. One light year is 9·4607 million million km (5·8 million million miles).

magnitude A measure of the brightness of a star or other astronomical object. The larger the number on the magnitude scale, the fainter the object. Objects brighter than zero magnitude have magnitudes that are negative numbers.

mantle A layer in the structure of a planet or satellite below the crust and overlying the core.

neutron An elementary particle with no electric charge. Together with protons, neutrons are found in the nuclei of atoms.

orbit The path one astronomical object takes around another. Closed orbits are usually elliptical: perfectly circular orbits rarely occur in nature. The orbits of some comets around the Sun appear to be open curves that are parabolic or hyperbolic in shape, though they could be just one end of a very large ellipse.

perigee The point in the orbit around the Earth of an artificial satellite or the Moon, that is closest to the Earth.

perihelion The point in the orbit of a planet (or other object) around the Sun, that is closest to the Sun.

poles The two points on a rotating body where its rotation axis intersects its surface, or the two opposite directions in space towards which an imaginary extension of the rotation axis points.

proton An elementary particle with a positive electric charge. Together with neutrons, protons are found in the nuclei of atoms.

red giant An evolved star that has expanded greatly in size and become relatively cool so that it appears red.

satellite An object in orbit around a larger body. The natural satellites of the planets are also known as moons.

solar wind The outward flow from the Sun of atomic particles, mainly electrons and protons, which stream through the Solar System and beyond.

white dwarf An old star that has used up its supply of hydrogen fuel and has collapsed to be much denser than ordinary matter.

zenith The point in the sky that is directly overhead an observer.

zodiac A belt of 12 constellations through which the Sun's path passes during the course of a year. The traditional zodiac constellations are Aries, Taurus, Gemini, Cancer, Leo, Virgo, Libra, Scorpius, Sagittarius, Capricornus, Aquarius and Pisces. Natural changes, and precise definitions of constellation boundaries mean that the Sun now actually passes through a thirteenth constellation, Ophiuchus.